Petroleum Engineering

The Springer series in Petroleum Engineering promotes and expedites the dissemination of new research results and tutorial views in the field of exploration and production. The series contains monographs, lecture notes, and edited volumes. The subject focus is on upstream petroleum engineering, and coverage extends to all theoretical and applied aspects of the field. Material on traditional drilling and more modern methods such as fracking is of interest, as are topics including but not limited to:

- Exploration
- Formation evaluation (well logging)
- Drilling
- Economics
- Reservoir simulation
- Reservoir engineering
- Well engineering
- Artificial lift systems
- Facilities engineering

Contributions to the series can be made by submitting a proposal to the responsible Springer Editor, Charlotte Cross at charlotte.cross@springer.com or the Academic Series Editor, Dr. Gbenga Oluyemi g.f.oluyemi@rgu.ac.uk.

More information about this series at http://www.springer.com/series/15095

Bodhisatwa Hazra · David A. Wood ·
Devleena Mani · Pradeep K. Singh ·
Ashok K. Singh

Evaluation of Shale Source Rocks and Reservoirs

 Springer

Bodhisatwa Hazra
Council of Scientific and Industrial
Research-Central Institute of Mining
and Fuel Research
Dhanbad, Jharkhand, India

Devleena Mani
University of Hyderabad
Hyderabad, Telangana, India

Ashok K. Singh
Council of Scientific and Industrial
Research-Central Institute of Mining
and Fuel Research
Dhanbad, Jharkhand, India

David A. Wood
DWA Energy
Bassingham, Lincolnshire, UK

Pradeep K. Singh
Council of Scientific and Industrial
Research-Central Institute of Mining
and Fuel Research
Dhanbad, Jharkhand, India

ISSN 2366-2646   ISSN 2366-2654   (electronic)
Petroleum Engineering
ISBN 978-3-030-13044-2   ISBN 978-3-030-13042-8   (eBook)
https://doi.org/10.1007/978-3-030-13042-8

This Springer imprint is published by the registered company Springer Nature Switzerland AG
The registered company address is: Gewerbestrasse 11, 6330 Cham, Switzerland

Organic-rich shales: more than just source rocks or reservoirs. The foundations on which the oil and gas industries persist.

Preface

Our research into organic-rich shales from various perspectives in recent years has led us to realize that although a substantial body of research built up over many decades, through painstaking experiments and analysis, that provides us with a fundamental basis of generic understanding and knowledge about these complex and versatile rock formations, there is still much to learn and reveal about them. Moreover, some of the analytical and interpretation techniques that are today routinely applied to organic-rich shales and considered to be well understood are in fact permeated with potential pitfalls and questionable assumptions making them prone to frequently provide misleading information that can and does lead to erroneous interpretations. Such errors can be costly and result in petroleum generation and production sweet spots being missed or their potential being poorly quantified in space and time and in terms of available resources.

Such a realization has stimulated us to compile this monograph to highlight how to extract the most useful information from the well-established analytical techniques applied to organic-rich shales. In order to achieve this, it is essential to avoid the pitfalls that can undermine data quality and reliability. In addition, it is necessary to recognize that some of the assumptions accepted as standard practice over decades, because they are easy to apply and/or simplify analysis/interpret now need to be questioned and scrutinized more closely. It is these requirements that determined the objectives and content of the six distinct chapters (plus introduction and conclusions) that constitute this work.

We describe in the introduction what is known with confidence about organic-rich shales, their variable characteristics, and duality as source rocks and reservoir rocks. Chapter 2 focuses on the fundamentals of source rock characterization and thermal maturity assessments and the uncertainties involved. Chapter 3 provides guidelines for the Rock-Eval pyrolysis techniques and explains how to avoid generating erroneous or ambiguous data. Chapter 4 highlights the frequent requirements to apply corrections for matrix retention effects when estimating the petroleum resource potential of organic-rich shales.

Chapter 5 addresses kerogen reaction kinetics and the timing and extent of kerogen conversion into petroleum that depends on the values used for key kinetic metrics. We highlight that some of the assumptions used by industry and academia for decades regarding distributions of kerogens' activation energies are inappropriate and lead to inaccurate estimates regarding the timing of kerogen conversion. More realistic kerogen kinetic assumptions lead to a more accurate model fits of Rock-Eval pyrogram S2 peaks. Chapter 6 describes the useful information, complementary to mineralogical analysis and Rock-Eval pyrolysis that is provided by geochemical biomarkers and isotopic analysis, helping to verify thermal maturity and pinpoint the most favourable petroleum production sweet spots in shales.

Chapter 7 considers and explains the complex nature of porosity, pore-size distributions and their fractal dimensions in organic-rich shales. The significance of such information to potential production rates and petroleum resource storage is considered, as are limitations of the various low-pressure gas adsorption techniques that potentially lead to data ambiguities and misinterpretation for some formations.

The key findings of Chaps. 1–7 are summarized in Chap. 8.

We hope that you enjoy this book and that it provokes you to think about shales more comprehensively, by integrating analysis from the various perspectives described. Having read and digested it, we additionally hope that you ultimately share our enthusiasm to conduct research that provides further insight and identifies new perspectives regarding the, as yet, only partially understood spectrum of rock formations identified as organic-rich shales.

Bassingham, UK   David A. Wood
March 2018

Acknowledgements

The Director, CSIR-Central Institute of Mining and Fuel Research, India, is acknowledged for granting permission to publish this work and also for providing the necessary infrastructure to carry out the work. The Director, CSIR-National Geophysical Research Institute, India, is also thankfully acknowledged for permitting the authors to access the necessary laboratory infrastructure. B. Hazra would additionally like to thank the Department of Science and Technology (DST; Ministry of Science and Technology, Government of India), for providing research funding through the DST-Inspire Faculty-Assured Opportunity of Research Career (AORC) scheme, which was partly used for this research work.

Contents

Chapter 1
Introduction

Organic-rich shales, until recently evaluated mainly as source-rocks feeding conventional oil and gas reservoirs, are now considered more broadly as both potential source rocks and as potential unconventional petroleum reservoirs. Laboratory analyses and models are now required to assess the extent of petroleum generation, and the quantities of petroleum retained within versus the extent of petroleum expelled from such formations over time. In this monograph we describe in detail the organo-petrographical properties, geochemical characteristics, and porous structures present within organic-rich shales. We highlight the analytical and interpretation protocols that need to be followed to provide meaningful assessment and characterization of these valuable formations.

Since the emergence of shales as self-contained petroleum systems, research focusing on their distinctive characteristics has increased (Schmoker 1995; Loucks et al. 2009, 2012; Psarras et al. 2017). Shales are fine-grained, siliciclastic sedimentary rocks, constituting about 50% of all sedimentary rocks (Boggs 2001). Often, the terms mudrocks, shales, and mudstones are used to represent the same rock types, although they are characteristically different (Peters et al. 2016). Shales are generally, finely bedded sedimentary units, which typically made up of many thin layers often deposited in cyclical sequences, each cycle with characteristic geochemistry (Slatt and Rodriguez 2012). Crucially some shales contain organic-matter from traces to significant contents, and, depending on their burial and thermal history this enables them to generate petroleum. Shales, together with coal reservoirs, oil sand deposits, and gas hydrate reservoirs, are categorized as unconventional petroleum reservoirs i.e. requiring some energy input to enable them to produce the petroleum held within them at commercially-viable rates (Peters et al. 2016; Wood and Hazra 2017a).

The fissility of shale is controlled by its organic matter content and the type of minerals present within it. Organic-rich shales consisting of platy or flaky minerals tend to be more fissile, while the presence of silicate and/or carbonate minerals in shale decreases their fissility. The overall mineralogy of shales tends to be complex as they mostly consist of fine-grained/silty products of abrasion, weathering-products (comprised mainly of residual clays) and chemical as well as biochemical additions (Pettijohn 1984). Carbonates and sulfides may arise as secondary burial cements

© Springer Nature Switzerland AG 2019
B. Hazra et al., *Evaluation of Shale Source Rocks and Reservoirs*,
Petroleum Engineering, https://doi.org/10.1007/978-3-030-13042-8_1

or replacement minerals in shales. Silicates such as feldspar, quartz, clays are often detrital (terrigenous), while some portion of these may also form together with a range of clay minerals during burial diagenesis (Boggs 2001). Many factors have an effect on the detailed composition and mineralogy of shales, viz. tectonic settings, provenance, depositional environments, grain size, and burial diagenesis. Illite, smectite, mixed layer illite/smectite, kaolinite, and chlorite are the dominant clay types that occur in shales (Chermak and Schreiber 2014). As the originally deposited sediments are subjected to increasing temperatures and pressures, they experience changes in their formation fluid chemistry, and consequently burial diagenesis occurs (Chermak and Schreiber 2014). The composition of clay minerals and their porous structure affects the petroleum storage capacity of shale reservoirs (Aringhieri 2004). Micropores present within the crystalline layers of clay minerals viz. kaolinite, illite etc., are known to provide adsorption sites for gases (Cheng and Huang 2004).

Trace element concentration in shales can also be significant and important. Generally, black shales are more enriched in trace elements compared to other shale types (Leventhal 1998). Moreover, some can also host economically relevant metal ores (Leventhal 1998). Trace element concentrations in shales can be used to identify paleoenvironmental conditions during shale deposition and paragenesis (Tribovillard et al. 2006).

Based on their organic matter content and Fe^{3+}/Fe^{2+} ratio, shales display different colours (Pettijohn 1984; Myrow 1990). Red coloration is essentially caused due to a higher Fe^{3+}/Fe^{2+} ratio, while green coloration is the result of more reducing conditions, i.e., lower Fe^{3+}/Fe^{2+} ratio (McBride 1974). Darker coloration of shales is typically related to higher organic matter content (Varma et al. 2014). In sedimentological studies of shales, the concept of 'shale facies' has been used extensively by different workers (Schieber 1989, 1990; Macquaker and Gawthorpe 1993). Composition-based 'shale facies' evaluation becomes more useful when evaluating depositional environments of those shaly-horizons which contain minimal sedimentary structures. For example, Macquaker and Gawthorpe (1993) identified five shaly-lithofacies based on clay contents, silt contents, biogenic contents, carbonate contents, and absence/presence of laminations within them. However, when present, primary structures including cyclical laminations, provide important clues on palaeo-depositional conditions of shales (Schieber 1998; Slatt and Rodriguez 2012). In shales, laminae are the most commonly observed sedimentary units showing different styles viz. continuous/even, discontinuous/uneven, lenticular, wrinkled etc. Each lamina represents a specific set of environmental factors operative contemporaneously with shale deposition. Internal lamina features within shales such as grading or preferential orientation of clay minerals can also provide important clues about the depositional environments (Schieber 1990). The microfabrics present within shales, can also provide necessary clues about processes operative during transportation and depositional phases (Schieber 1998). However, the high amount of compaction other diagenetic induced changes that muds undergo after deposition, raises some queries about the usage of microfabrics for reconstructing their depositional environmental.

The shale-lithofacies concepts, identifying the lithological characteristics of these formations in terms of organic richness, brittleness, mineral composition, have also emerged as important tool for shale gas and oil exploration and development (Dill et al. 2005; Tang et al. 2016). Since organic-rich shales are mainly characterized by ultra-low permeabilities (Guarnone et al. 2012), the proportion of clay minerals determines the shale 'plasticity' or 'fracability' (Jarvie et al. 2007; Tang et al. 2016). Generally, shales are considered to be brittle when they consist of <40% clay minerals (Tang et al. 2016), and such conditions are considered to be favorable for hydraulic fracturing. A dominance of quartz and carbonate minerals over clay minerals generally leads to an easier initiation and propagation of fractures created during fracture stimulation (Wood and Hazra 2017c). Clay-rich shales, on the other hand, possess self-sealing characteristics that inhibit the propagation of induced fractures (Josh et al. 2012). The importance of shale-lithofacies is emphasized by Tang et al. (2016) in studies of the marine shales belonging to the Longmaxi formation (Silurian) from Sichuan basin, China. For organic-rich siliceous shale horizons, they observed enhanced free gas storage space, while for the organic-rich argillaceous shale horizons they observed enhanced methane sorption capacity. This allowed them to conclude that the shale-lithofacies identified for these two horizons can be used as exploration tools to reveal potential 'pay zones'.

Determining the ability of shale to generate petroleum, what fraction of that petroleum it expels (defining its source rock properties) and what fraction of that petroleum it retains (defining is unconventional reservoir rock properties) is a priority in characterizing shales and their petroleum prospectivity. Their prospectivity is strongly dependent upon the nature, type, and richness of the organic-matter present within them, the composition of the organic-matter and its thermal maturity level (Wood and Hazra 2017b). The potential of organic-matter to generate petroleum depends on whether the carbon of organic-matter is associated with hydrogen or not. The more hydrogen present, the greater the quantity of petroleum fluids generated is likely to be (Dembicki 2009). The petroleum present within shales maybe localized in an adsorbed state both within the porous structures of organic-matter and other mineral matter (Curtis 2002; Ross and Bustin 2009). Petroleum components may also exist in free form or in dissolved form. However, the adsorbed gas phase present within the shale-pores, comprises the major component, and thus highlights the importance of mapping of pore structural facets of shale reservoirs on macro, micro and nano scales. Laboratory characterization of shales in terms of the above-mentioned attributes is of paramount significance for assessing their potential as petroleum reservoirs.

Geochemical characterization of the organic matter contained in organic-rich shales and its capacity to generate, retain, and expel petroleum provides insight to their potential as gas and/or oil reservoirs (Jarvie et al. 2007). Open system programmed pyrolysis techniques coupled with organic petrologic data derived from light microscopy, have emerged as useful tools for geochemical characterization of organic-rich shales (Carvajal-Ortiz and Gentzis 2015; Hackley and Cardott 2016; Romero-Sarmiento et al. 2016; Hazra et al. 2017). Low-pressure gas adsorption techniques reveal useful insight to the morphology of porous structures and pore size distributions in such formations. However, that information requires careful interpretation in order to gain a meaningful representation of porosity at the nanoscale.

In this monograph we provide a step-by-step account of the different facets of organic-matter characterization in shales. We build on established techniques (e.g., organic petrology, geochemistry, source rock evaluation, and low-pressure gas adsorption techniques) and show how they have been refined and more extensively exploited in recent years to provide extensive insight to the petroleum prospectivity of shales. The different analytical protocols that need to be followed for proper shale source-rock property evaluation are explained. Further, we establish how the reaction kinetics of petroleum formation, and the presence of different biomarkers within shales can provide of when petroleum generation is likely to have occurred and how much of the kerogen has already been converted to petroleum versus that not yet converted.

In Chap. 2 we discuss the fundamentals of source-rock geochemistry comprising of organic-matter richness, types, and maturity, which determines their hydrocarbon generation potential. Vitrinite reflectance in shales is used to assess their thermal maturity levels, however, the quality of vitrinite grains should be carefully monitored, and as in shales they are inherently low in contents (dispersed) and often show mottled/pitted surfaces. In Chap. 3, we address the most widely used pyrolysis technique (Rock-Eval) for mapping shale petroleum generation parameters. We provide key guideline for using Rock-Eval data to generate 'error-free' source-rock assessment. Several factors viz. particle crush-size, FID signal, S2 pyrogram shape, FID linearity, S4CO$_2$ oxidation graphics, have the potential to distort source-rock characterization using Rock-Eval if these guidelines are disregarded. In Chap. 4, we discuss the impact of shale-matrix on retention of hydrocarbons during open-system anhydrous pyrolysis experiments such as the Rock-Eval technique. The role of both kerogen type and rock-matrix on the elution of petroleum-related fluids from shale during pyrolysis is considered. Correction of matrix-retention and inert organic-matter effects are shown to be important, especially when estimating the petroleum potential of a shale reservoir. In Chap. 5, we present details about the conversion of kerogen to petroleum fluids and the reaction kinetics associated with that conversion. Data measurement, analysis, modeling and interpretation techniques associated with kerogen's conversion in organic-rich sediments are reviewed. Chapter 6 presents the occurrence, significance, and importance of biomarkers and isotopes, and the important information they provide regarding the provenance of the sediments, depositional environment, and thermal maturity of organic-matter in shales. Chapter 7 addresses the detailed analysis of porosity and pore size distribution of organic-rich shales. Using low pressure gas adsorption techniques, we characterize the pore size distribution attributes of shales as a function of particle crush-size, organic richness, and thermal maturity level. We further discuss the presence and significance of fractal dimensions in the porosity of shales and how that influences their petroleum storage potential.

References

Aringhieri R (2004) Nanoporosity characteristics of some natural clay minerals and soils. Clays Clay Miner 52:700–704

Boggs Jr S (2001) Sedimentary structures. In: Principles of sedimentology and stratigraphy, 3rd edn. Prentice-Hall, Upper Saddle River (New Jersey), pp 88–130

Carvajal-Ortiz H, Gentzis T (2015) Critical considerations when assessing hydrocarbon plays using Rock-Eval pyrolysis and organic petrology data: data quality revisited. Int J Coal Geol 152:113–122

Cheng AL, Huang WL (2004) Selective adsorption of hydrocarbon gases on clays and organic matter. Org Geochem 35:413–423

Chermak JA, Schreiber ME (2014) Mineralogy and trace element geochemistry of gas shales in the United States: environmental implications. Int J Coal Geol 126:32–44

Curtis JB (2002) Fractured shale-gas systems. AAPG Bull 86:1921–1938

Dembicki H Jr (2009) Three common source rock evaluation errors made by geologists during prospect or play appraisals. AAPG Bull 93:341–356

Dill HG, Ludwig RR, Kathewera A, Mwenelupembe J (2005) A lithofacies terrain model for the Blantyre Region: implications for the interpretation of palaeosavanna depositional systems and for environmental geology and economic geology in southern Malawi. J Afr Earth Sci 41(5):341–393

Guarnone M, Rossi F, Negri E, Grassi C, Genazzi D, Zennaro R (2012) An unconventional mindset for shale gas surface facilities. J Nat Gas Sci Eng 6:14–23

Hackley PC, Cardott BJ (2016) Application of organic petrography in North American shale petroleum systems. Int J Coal Geol 163:8–51

Hazra B, Dutta S, Kumar S (2017) TOC calculation of organic matter rich sediments using Rock-Eval pyrolysis: critical consideration and insights. Int J Coal Geol 169:106–115

Jarvie DM, Hill RJ, Ruble TE, Pollastro RM (2007) Unconventional shale-gas systems: The Mississippian Barnett Shale of north-central Texas as one model for thermogenic shale-gas assessment. AAPG Bull 91(4):475–500

Josh M, Esteban L, Piane CD, Sarout J, Dewhurst DN, Clennell MB (2012) Laboratory characterisation of shale properties. J Petrol Sci Eng 88–89:107–124

Leventhal JS (1998) Metal-rich black shales: formation, economic geology and environmental considerations. In: Schieber J, Zimmerle W, Sethi P (eds) Shales and mudstones II. E. Schweizerbart'sche Verlagsbuchhandlung, Stuttgart

Loucks RG, Reed RM, Ruppel SC, Jarvie DM (2009) Morphology, genesis, and distribution of nanometer-scale pores in siliceous mudstones of the Mississippian Barnett Shale. J Sediment Res 79:848–861

Loucks RG, Reed RM, Ruppel SC, Hammes U (2012) Spectrum of pore types and networks in mudrocks and a descriptive classification for matrix-related mudrock pores. AAPG Bull 96:1071–1098

MacQuaker JHS, Gawthorpe RL (1993) Mudstone lithofacies in the Kimmeridge clay formation, Wessex Basin, Southern England: implications for the origin and controls of the distribution of mudstones. J Sediment Petrol 63:1129–1143

McBride EF (1974) Significance of color in red, green, purple, olive, brown and gray beds of Difunta Group, northeastern Mexico. J Sediment Petrol 44:760–773

Myrow PM (1990) A new graph for understanding colors of mudrocks and shales. J Geol Educ 38:16–20

Peters KE, Xia X, Pomerantz AE, Mullins OC (2016) Geochemistry applied to evaluation of unconventional resources. In: Ma YZ, Holditch SA (eds) Unconventional oil and gas resources handbook: evaluation and development. Gulf Professional Publishing, Waltham, MA, pp 71–126

Pettijohn FJ (1984) Sedimentary rocks. Harper & Row, New York

Psarras P, Holmes R, Vishal V, Wilcox J (2017) Methane and CO_2 adsorption capacities of kerogen in the Eagle Ford shale from molecular simulation. Accounts Chem Res 50(8):1818–1828

Romero-Sarmiento M-F, Pillot D, Letort G, Lamoureux-Var V, Beaumont V, Huc A-Y, Garcia B (2016) New Rock-Eval method for characterization of unconventional shale resource systems. Oil & Gas Science and Technology 71:37

Ross DJK, Bustin RM (2009) The importance of shale composition and pore structure upon gas storage potential of shale gas reservoirs. Mar Pet Geol 26:916–927

Schieber J (1989) Facies and origin of shales from the Mid-Proterozoic Newland Formation, Belt Basin, Montana, USA. Sedimentology 36:203–219

Schieber J (1990) Significance of styles of epicontinental shale sedimentation in the Belt basin, Mid-Proterozoic of Montana, U.S.A. Sed Geol 69:297–312

Schieber J (1998) Deposition of mudstones and shales: overviews, problems, and challenges. In: Schieber J, Zimmerle W, Sethi P (eds) Mudstones and shales (vol 1). Characteristics at the basin scale. Schweizerbart'sche Verlagsbuchhandlung, Stuttgart

Schmoker JW (1995) Method for assessing continuous-type (unconventional) hydrocarbon accumulations. In: Gautier DL, Dolton DL, Takahashi KI, Varnes KL (eds) National assessment of United States oil and gas resources—results, methodology, and supporting data: U.S. Geological Survey Digital Data Series 30, CD-ROM

Slatt RM, Rodriguez ND (2012) Comparative sequence stratigraphy and organic geochemistry of gas shales: commonality or coincidence? J Nat Gas Sci Eng 8:68–84

Tang X, Jiang Z, Huang H, Jiang S, Yang L, Xiong F, Chen L, Feng J (2016) Lithofacies characteristics and its effect on gas storage of the Silurian Longmaxi marine shale in the southeast Sichuan Basin. China J Nat Gas Sci Eng 28:338–346

Tribovillard N, Algeo TJ, Lyons T, Riboulleau A (2006) Trace metals as paleoredox and paleoproductivity proxies: an update. Chem Geol 232:12–32

Varma AK, Hazra B, Srivastava A (2014) Estimation of total organic carbon in shales through color manifestations. J Nat Gas Sci Eng 18:53–57

Wood DA, Hazra B (2017a) Characterization of organic-rich shales for petroleum exploration & exploitation: a review—Part 1: Bulk properties, multi-scale geometry and gas adsorption. J Earth Sci 28(5):739–757

Wood DA, Hazra B (2017b) Characterization of organic-rich shales for petroleum exploration & exploitation: a review—Part 2: Geochemistry, thermal maturity, isotopes and biomarkers. J Earth Sci 28(5):758–778

Wood DA, Hazra B (2017c) Characterization of organic-rich shales for petroleum exploration & exploitation: a review—Part 3: Applied geomechanics, petrophysics and reservoir modeling. J Earth Sci 28(5):779–803

Chapter 2
Source-Rock Geochemistry: Organic Content, Type, and Maturity

One of the first and perhaps the most vital segment for probing hydrocarbon plays is the source-rock appraisal or geochemical screening (Jarvie 2012a, b). A successful petroleum generating source-rock must fulfill with respect to its organic-matter amount, type, and thermal maturity (Tissot and Welte 1978). For shale or coal systems that integrate source rock and reservoir in the same formation (unconventional plays), geochemical screening needs to cover several variables, including: the amount of free oil/gas present, the amount and quality of organic-matter present, thermal maturity level, proportion of reactive and residual/non-reactive carbon. These variables collectively determine the geochemical 'quality' of an organic rich formation and are usefully summarized as a geochemical profile.

2.1 Organic Richness

The organic-matter contents of potential petroleum source rocks and/or unconventional reservoirs influence their petroleum generation capabilities in terms of quantity and the petroleum capacity that can be stored within their matrices. Rock-Eval pyrolysis (discussed in details in Sect. 2.2), an open-system programmed-pyrolysis mechanism, is widely used for geochemical profiling, in particular providing an indication of organic content, by measuring total organic carbon (TOC) and distinguishing its components, i.e., the free oil/gas present, the heavier pyrolyzate fraction present, reactive and residual carbon content. Opinions vary regarding the minimum organic-matter or TOC requirement/threshold, for a formation to be designated as a potential source-rock or viable petroleum generating system. Welte (1965) proposed that a prospective source-rock should have at least 0.5 wt% TOC, while Peters and Cassa (1994) classified source-rocks into different categories based on their TOC content, as follows: poor (0–0.5 wt% TOC), fair (0.5–1 wt% TOC), good (1–2 wt% TOC), very good (2–4 wt% TOC), and excellent (>4 wt% TOC). Others have attempted to justify slightly different TOC thresholds. For instance, Jarvie and Lundell (1991), Bowker (2007), Burnaman et al. (2009) suggested different minimum TOC require-

© Springer Nature Switzerland AG 2019
B. Hazra et al., *Evaluation of Shale Source Rocks and Reservoirs*,
Petroleum Engineering, https://doi.org/10.1007/978-3-030-13042-8_2

ments for unconventional shale reservoirs. There are several factors influencing the uncertainty regarding the minimum TOC requirement of a potential source-rock, viz. kerogen types present, the thermal maturity levels of the organic-matter, and host-rock mineralogy (Wood and Hazra 2017). The influence of kerogen type, host-rock mineralogy and thermal maturity levels on TOC are discussed in subsequent sections.

2.2 Organic Composition

The quantity of organic carbon does not on its own determine a source-rock's petroleum generation potential. The amount of hydrogen present within the organic-matter has a significant influence on the amount of petroleum that can be generated. Organic carbon can be generative or non-generative, depending on the type of organic-matter present, and its thermal maturity levels. Kerogens are the most significant type of organic carbon present in shales. Different kerogen types have different petroleum generation and expulsion potential. Kerogens display a range of distinctive elemental compositions and are traditionally classified into 4 types—types I, II, III, and IV (van Krevelen 1961, 1993). Some kerogens are more prone to generating liquid petroleum, others are more prone to generating natural gas, and some generate a mixture of liquid petroleum and gas, while some remain incapable of generating significant petroleum quantities at all. These distinguishing petroleum-generating capabilities of the kerogen types are fundamentally linked to their hydrogen content when in a thermally immature state.

Types I and II kerogen are characterized by higher initial elemental hydrogen content, higher atomic H/C ratios, and lower atomic O/C ratios, in comparison to types III and IV kerogen. The higher aliphatic contents of types I and II kerogen, owing to their higher hydrogen contents, enable them to generate greater quantities of oil, in contrast to the hydrogen-poor types III and IV kerogen. Type III kerogen is marked by lower H/C ratios, lower aliphatic contents, and higher atomic O/C ratios compared to types I and II kerogen, are prone to generate dry-gas as they become thermally mature. Type IV kerogen on the other hand, is marked by extremely low hydrogen content, and variable atomic O/C ratios, making it inert in terms of petroleum generation potential. Type II kerogens contain more cyclic-aliphatic-structures (naphthenic rings) than Type I kerogens. This promotes the generation of naphthenic-oil from Type II kerogens. On the other hand, more waxy, paraffinic oils are typically generated from Type I kerogens (owing to their long-chained aliphatic-structures).

Type II kerogens are typically the dominant kerogen-type in shale-plays yielding liquid petroleum products (Romero-Sarmiento et al. 2014). On the other hand, Hackley and Cardott (2016) identified that solid bitumen is the most dominant petrographic constituent with organic content in North American mature shale-gas formations. Petroleum geochemists' define bitumen as that fraction of sedimentary organic matter which may be extracted using common organic solvents, while the non-extractable fraction is referred to as kerogen. However, the solid bitumen referred to by Hackley and Cardott (2016) represents macerals (microscopically identifiable organic constituents of organic-sedimentary rocks) formed secondarily as part of the petroleum generation and modification processes occurring during thermal maturation.

The most common types of kerogen macerals that are present in organic-rich sedimentary rocks are vitrinites, inertinites, and liptinites (including alginites). Vitrinite macerals, in shales and other organic-rich sedimentary rocks represent those microscopic constituents which originate from the woody-materials consisting of the cellulose and lignin remains of vascular-plants. Inertinite macerals, on the other hand, typically represent those organic constituents that have undergone (partial or complete) combustion and oxidation (either prior to burial or in the early stages of burial). The liptinite macerals consist primarily of amorphous organic-matter originating from algal or bacterial materials. Shales deposited in lacustrine environments typically contain liptinites made up of chemically discrete parts of plants such as spores and cuticles (Hackley and Cardott 2016).

In fine-grained shales the sub-macerals vitrinite and inertinite groups are generally indistinguishable and are simply referred to as vitrinites and inertinites (Hackley and Cardott 2016). However, in the case of some coal-measures associated with interbedded shales, the sub-maceral groups of vitrinites and inertinites can be distinguished. Figure 2.1 shows a photomicrograph of a high-TOC shale sample, under oil immersion, from a borehole penetrating an interbedded coal and shale sequence from the eastern Raniganj basin, India. The vitrinite sub-maceral collotelinite (unstructured, homogeneous) is clearly discernible in the sample. Vitrinite reflectance analysis is a standard technique for assessing the maximum thermal maturity level reached by specific coal and shale formations. Such analysis is conducted typically on collotelinite macerals, as they are homogeneous, and show increasing reflectivity with increasing thermal-maturity levels.

The vitrinite group of macerals, under reflected white-light (oil immersion) is characterized by a medium grey colour. They tend to possess intermediate reflectances between the darker liptinites and lighter inertinites (Fig. 2.2). The liptinites, although they appear dark under reflected white-light generate the least reflectance among the three maceral groups. However, liptinite shows fluorescence of different intensities when energized with short-wavelength radiation, which is not showed by the other groups of macerals. On the other hand, inertinites, are characterized by the highest reflectance among the three groups of macerals.

Rarely does any source-rock consist of only a single kerogen-type, rather admixed macerals and kerogen-types tend to predominate (Fig. 2.2). A key limitation of any bulk geochemical analysis (elemental or Rock-Eval) for either coals or shales, is that it represents the composite effect of all the organic-matter types (macerals) present, smoothing and sometimes obscuring their distinctive characteristics and petroleum-generation capabilities. Identification and consideration of the relative concentrations of the different organic matter present and their respective levels of thermal maturity enables more rigorous petroleum generation modeling and interpretation. Further, thorough inspection of macerals allows organic petrologists to establish the extent to which the component kerogens have been converted into petroleum products. For

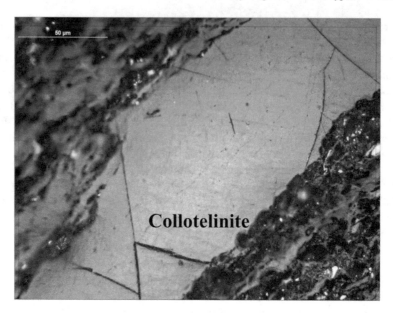

Fig. 2.1 Photomicrograph of Upper Permian Raniganj Formation shale from Raniganj basin, India, with TOC content of 9.44 wt%, showing presence of unstructured collotelinite maceral (vitrinite group) at peak thermal maturity

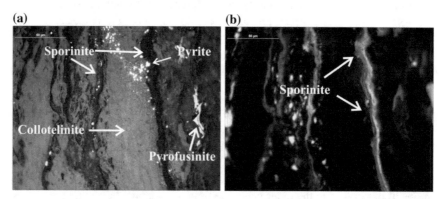

Fig. 2.2 Photomicrograph of an Upper Permian Raniganj Formation carbonaceous-shale from Raniganj basin, India, with TOC content of 26.56 wt% at peak thermal maturity. These images reveal the presence of unstructured collotelinite maceral (vitrinite group), pyrofusinite (inertinite group), sporinite (liptinite group), and pyrite (highly reflecting mineral matter in **a**). **a** image taken under reflected white light; **b** image taken under blue-light florescence. Note that only sporinite maceral shows florescence under blue-light florescence

example, Fig. 2.3 shows the presence of micrinite, along with vitrinite and sporinite (distinguished by its fluorescence—dark under reflected white light), in a high-TOC (7.04 wt%) Upper Permian Raniganj Formation shale from India. The presence of the highly reflecting micrinite maceral in Fig. 2.3 indicates that some hydrocarbon expulsion has taken place from the sample. Micrinite occurs mainly as secondary maceral generated during the transformation of hydrogenated organic-matter into petroleum-like substances. Micrinites are thus often found to occur in association of liptinite macerals, and are interpreted to represent the residue after the conversion of liptinites into petroleum products (Taylor et al. 1998). Gorbanenko and Ligouis (2014) in their study on early mature and postmature Posidonia Shale from NW Germany, observed marked differences in the properties of telalginite maceral (alginite) with increasing maturity. For a few mature shales, the telaginites display 'corroded' or 'pitted' surfaces indicating generation of petroleum from them. For the postmature shales on the other hand, fluorescence properties of the liptinite macerals are no longer present. These postmature shales contain micrinites within the groundmass together with secondarily formed carbonate minerals, pyrite, and pyrolytic carbon. Also, the primary vitrinites of these post-mature shales display distinct fissures/pores, which are not present in the less mature samples. Petroleum fluids generated from the liptinite macerals are likely responsible for these observed changes in the morphology of their organic minerals. Detailed petrographic analysis of the organic macerals therefore helps the analysts in the initial petroleum-generation potential of organic-rich shales.

Reactivity of macerals is of interest for coal geologists dealing with technological applications and utilization of coal. Initially, the inertinites were considered to be completely inert and infusible for all technological characterization purposes (Stach

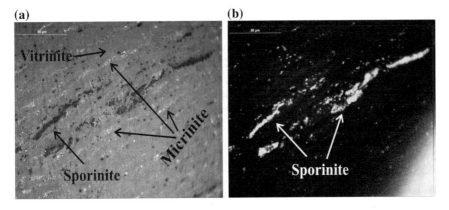

Fig. 2.3 Photomicrograph of an Upper Permian Raniganj Formation high-TOC shale (TOC: 26.56 wt%) from Raniganj basin, India, at the early stage of thermal maturity. It reveals the presence of vitrinite, sporinite, and micrinite. **a** image under reflected white light; **b** same-view of image under blue-light florescence. The photomicrographs indicate the close association of micrinite with spronite in these samples

1952). Subsequent studies by Ammosov et al. (1957), Košina and Heppner (1985), Rentel (1987), Varma (1996, 2002) revealed some reactivity associated with inertinites and their positive roles during technological applications such as enhancing coal carbonization and combustion. Further, Hazra et al. (2015) identified some positive reactivity of inertinite macerals influencing petroleum generation from Permian shales in India, but to a smaller extent than the other maceral groups (mainly type III) present in those shales.

2.3 Organic Maturity

The thermal maturity level of organic-matter within shales and coals is essential in understanding their petroleum-generation potential. During the process of sedimentation and concomitant burial, the bulk properties of organic-matter or kerogen with shales evolve thermally (Tissot and Welte 1978). As the thermal maturity of kerogen increases it is not only partially transformed into petroleum fluids but also, significantly, that process is accompanied with the formation of secondary porosity within the kerogen itself (Loucks et al. 2012). Porosity within kerogen constitutes an important component of the petroleum storage capacity for shales and other unconventional organic-rich reservoirs.

During the initial diagenetic stage of burial, there is marked loss in oxygen content from the organic-matter resulting in strong decrease in atomic O/C ratio, accompanied by only small changes in atomic H/C ratios. Kerogen at this stage is thermally immature, meaning that it is incapable of generating oil. It is during the catagenesis stage, that petroleum fluids are principally generated from organic-matter, resulting in the decrease of H/C ratios as the kerogens advance through stages of thermal maturity.

The most commonly used tool to determine the maturity of kerogen utilizes the optical microscopic technique to determine the reflectance of the vitrinite maceral (collotelinite; type III kerogen) measured under reflected lights (Teichmüller 1987; Mukhopadhyay and Dow 1994; Taylor et al. 1998). Vitrinite reflectance measurement (expressed as a percentage and referred to by the abbreviation Ro %) is not only a robust technique for assessing kerogen thermal maturity it is also relatively cheap and easy to conduct. The majority of studies focused on establishing the petroleum-generation potential of shales use vitrinite reflectance measurements to determine thermal maturity (Curtis et al. 2012; Hazra et al. 2015; Hackley and Cardott 2016).

Based on characterization of a suite of Barnett Shales (Texas, U.S.A), Jarvie et al. (2005) provided thermal maturity-profiling of immature, oil-window mature, condensate wet-gas mature, dry-gas mature shales, with Ro values of <0.55%, 0.55–1.15%, 1.15–1.40%, >1.40%, respectively. Generally, from H-rich kerogen (type I and type II), oil is liberated within vitrinite reflectance values of 0.6–1.3%, while gas is liberated from type III kerogen (e.g. vitrinite) or due to secondary oil to gas cracking at ≥1.0% reflectance values (Hunt 1996). However, some kerogens have been shown to generate petroleum at lower thermal maturity levels (Ro ≤ 0.40%), principally due to the presence of sulphur within the kerogen or their inherent high oil-yielding character. The presence of reactive-components in some kerogens enables their petroleum-generating reactions to initiate at relatively low thermal maturities (Lewan 1998; Lewan and Ruble 2002).

Baskin and Peters (1992) observed that for Miocene Monterey Formation in California, the kerogen was typically rich in chemically-bound sulphur (generally >10 wt%), which enabled liberation of sulphur-rich oil/petroleum at lower thermal maturities. The carbon-sulphur bonds in high sulphur type-II kerogens cleave readily enabling hydrocarbon molecular reactions to occur at much lower temperatures compared to carbon-carbon or other bonds (Lewan 1985; Tissot et al. 1987). It is apparent that the reaction rates of the kerogen-petroleum generating processes can be varied in relation to the thermal maturity level of formations by the structure and chemical components incorporated within the organic macerals. Lewan (1998), using pyrolysis experiments, documented that rather than weakness of carbon-sulphur bonds, the presence of sulphur radicals at the initial maturation stages, control formation of petroleum at lower thermal maturity levels. Under some circumstances, the vitrinite reflectance values within a shale can be suppressed, yielding inaccurate indications of thermal maturity. One cause of this can be oil impregnation of the vitrinite emanating from the associated hydrogen-rich kerogens in a mixed-kerogen source rock. For example, Kalkreuth (1982), Goodarzi et al. (1994) documented reflectance suppression of vitrinites in coals containing significant liptinite contents. For some Australian oil shale samples, reduced vitrinite reflectance values with increasing alginite contents were observed by Hutton and Cook (1980). Despite this notion being forwarded by many, the plausible causes and mechanisms of reflectance suppression of vitrinites in presence of liptinites, still remains somewhat controversial (Peters et al. 2018). Few authors have interpreted that suppression in vitrinite reflectances may be caused due to the impact of depositional settings. Newman and Newman (1982) observed marked difference in reflectance values of two New Zealand coals, having near similar calorific value and moisture yield. They opined that the suppression in reflectance value for one sample was caused due to redox potential in the depositional environment, and not due to liptinite content.

Beyond the principal zone of oil generation (initial catagenesis zone, Ro: 0.60–1.3%), is the condensate to wet-gas generation window (late catagenesis zone, Ro: 1.3–2%), and dry gas zone (metagenesis zone, Ro: >2%). Ideally, with increasing depths in a geological basin, the vitrinite reflectance values of shales and coals increase, with concomitant increases in carbon content and aromaticity, and decreases in hydrogen and oxygen contents, and lowering of aliphaticity of organic-matter (Tissot and Welte 1978). In case of major geological events, such as igneous intrusions and thrust folding, locally, the measured values of vitrinite reflectance in depositional sequences of shales and coals can be inverted. The effect of igneous intrusions, on shale petroleum-generation properties, pore structural attributes, gas sorption capacity, and vitrinite reflectance was documented by Hazra et al. (2015). Shales in close proximity to igneous intrusions display enhanced vitrinite reflectance values compared to the shales more distal from the intrusion above and below the thermally-affected zone. While petroleum fluids are driven-off from metamorphosed zone in the immediate vicinity of the intrusion, the shale immediately adjacent to that metamorphosed zone experiences elevated temperatures and pressures that enhance their thermal maturity to some degree. Some of those heat-affected shales may form horizons with potentially exploitable natural gas. Due to the impact of intrusion, hydrocarbon volatiles are liberated and expelled from the organic-matter, resulting in increasing their aromaticity, lowering their aliphaticity, formation of devalotilization vacuoles, vesicles, cracks, and fissures (Singh et al. 2007; Hazra et al. 2015). Figure 2.4 shows a photomicrograph of an igneous intrusion-affected shale from the Raniganj Basin, India. It reveals the development of bireflectance within the vitrinite grains caused mainly due to the impact of temperature and pressure shock induced by the igneous intrusion. Bireflectance is a measure of the anisotropy of vitrinite reflectance caused during coalification under the effect of directional variations of temperature and pressure (Levine and Davis 1989). Often the thermally metamorphosed organic-rich rocks, show development of natural 'chars' or natural 'cokes', depending on the inherent character of the organic-matter i.e. whether they are coking or non-coking (Singh et al. 2008).

The use of vitrinite reflectance as a measure of thermal maturity has several advantages, including: the presence of vitrinite in shales deposited since Palaeozoic times and the ability to generate reproducible measurements inexpensively with relatively simple laboratory equipment. It is typically much more expensive and time consuming to use geochemical biomarkers to measure thermal maturity than vitrinite reflectance. Nevertheless, vitrinite reflectance measurements in shales can some-

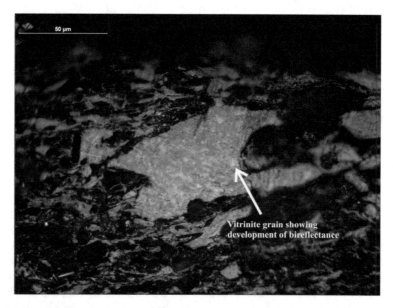

Fig. 2.4 Photomicrograph of a heat-affected Lower Permian Barakar Formation shale from Raniganj basin, India, showing the development of bireflectance in vitrinite (Hazra et al. 2015)

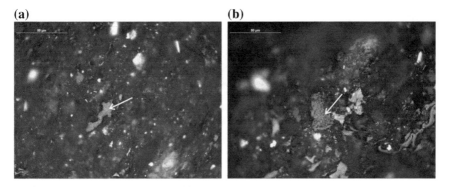

Fig. 2.5 Photomicrograph of a Lower Permian Barren Measures shale from the Raniganj basin, India, highlighting with the white arrow the presence of a vitrinite grain with **a** a pitted surface and, **b** a mottled surface. Vitrinite grains such as these, yield erroneous vitrinite reflectance measurements, owing to the absence of a smooth reflecting surface

times provide misleading or erroneous results. Vitrinite grains in shales, owing to their dispersed nature, are generally less abundant and smaller in size than those present in coals. Further, the vitrinite grains can often be oxidized, altered, with their surfaces pitted, making them error-prone when used to provide reflectance measurements. Moreover, some shales contain abundant reworked vitrinite eroded from older formations and mixed with contemporaneous vitrinite in the formation. Figure 2.5 shows some examples on dispersed vitrinite grains in the Lower Permian Barren Measures shales from India that yield erroneous vitrinite reflectance measurements. Under such circumstances, analysts should use other proxies for thermal maturity measurements, such as geochemical biomarkers and/or Rock-Eval T_{max} (discussed in detail in Sect. 2.2) for providing consistent estimates of thermal maturity.

References

Ammosov II, Eremin IV, Suchenko SI, Oshurkova IS (1957) Calculation of coking charges on the basis of petrographic characteristics of coals. Koks Khim 12:9–12 (in Russian)

Baskin DK, Peters KE (1992) Early generation characteristics of sulfur-rich Monterey kerogen. Am Assoc Pet Geol Bull 76:1–13

Bowker KA (2007) Barnett shale gas production, Fort Worth Basin: issues and discussion. AAPG Bull 91(4):523–533

Burnaman MD, Xia WW, Shelton J (2009) Shale gas play screening and evaluation criteria. China Pet Explor 14(3):51–64

Curtis ME, Cardott BJ, Sondergeld CH, Rai CS (2012) Development of organic porosity in the Woodford Shale with increasing thermal maturity. Int J Coal Geol 103:26–31

Goodarzi F, Snowdon L, Gentzis T, Pearson D (1994) Petrological and chemical characteristics of liptinite-rich coals from Alberta, Canada. Mar Pet Geol 11:307–319

Gorbanenko OO, Ligouis B (2014) Changes in optical properties of liptinite macerals from early mature to post mature stage in Posidonia Shale (Lower Toarcian, NW Germany). Int J Coal Geol 133:47–59

Hackley PC, Cardott BJ (2016) Application of organic petrography in North American shale petroleum systems. Int J Coal Geol 163:8–51

Hazra B, Varma AK, Bandopadhyay AK, Mendhe VA, Singh BD, Saxena VK, Samad SK, Mishra DK (2015) Petrographic insights of organic matter conversion of Raniganj basin shales, India. Int J Coal Geol 150–151:193–209

Hunt JM (1996) Petroleum geochemistry and geology. W.H. Freeman and Company, New York

Hutton AC, Cook AC (1980) Influence of alginite on the reflectance of vitrinite from Joadja, NSW, and some other coals and oil shales containing alginite. Fuel 59:711–714

Jarvie DM, Lundell LL (1991) Hydrocarbon generation modeling of naturally and artificially matured Barnett Shale, Fort Worth Basin, Texas. In: Southwest Regional Geochemistry Meeting, September 8–9, 1991, The Woodlands, Texas, 1991. http://www.humble-inc.com/Jarvie_Lundell_1991.pdf

Jarvie DM, Hill RJ, Pollastro RM (2005) Assessment of the gas potential and yields from shales: the Barnett Shale model. In: Cardott BJ (ed) Unconventional energy resources in the southern midcontinent, 2004 symposium. Oklahoma Geological Survey Circular, vol 110, pp 37–50

Jarvie DM (2012a) Shale resource systems for oil and gas: part 1—shale–gas resource systems. In: Breyer JA (ed) Shale reservoirs—giant resources for the 21st century. AAPG Memoir 97, pp 69–87

Jarvie DM (2012b) Shale resource systems for oil and gas: part 2—shale–oil resource systems. In: Breyer JA (ed) Shale reservoirs—giant resources for the 21st century. AAPG Memoir 97, pp 89–119

Kalkreuth WD (1982) Rank and petrographic composition of selected Jurassic-Lower Cretaceous coals of British Columbia, Canada. Can Petrol Geol Bull 30:112–139

Kôŝina M, Heppner P (1985) Macerals in bituminous coals and the coking process, 2. Coal mass properties and the coke mechanical properties. Fuel 64:53–58

Levine JR, Davis A (1989) Reflectance anisotropy of Upper Carboniferous coals in the Appalachian foreland basin, Pennsylvania, U.S.A. In: Lyons PC, Alpern B (ed) Coal: classification, coalification, mineralogy, trace-element chemistry, and oil and gas potential. Int J Coal Geol 13:341–374

Lewan MD (1985) Evaluation of petroleum generation by hydrous pyrolysis. Phil Trans R Soc Lond A 315:123–134

Lewan MD (1998) Sulphur-radical control on petroleum formation rates. Nature 391:164–166

Lewan MD, Ruble TE (2002) Comparison of petroleum generation kinetics by isothermal hydrous and non-isothermal open-system pyrolysis. Org Geochem 33:1457–1475

Loucks RG, Reed RM, Ruppel SC, Hammes U (2012) Spectrum of pore types and networks in mudrocks and a descriptive classification for matrix-related mudrock pores. AAPG Bull 96:1071–1098

Mukhopadhyay PK, Dow WG (eds) (1994) Vitrinite reflectance as a maturity parameter: applications and limitations. ACS Symposium Series 570, pp 294

Newman J, Newman NA (1982) Reflectance anomalies in Pike River coals: evidence of variability of vitrinite type, with implications for maturation studies and "Suggate rank". NZ J Geol Geophys 25:233–243

Peters KE, Cassa MR (1994) Applied source rock geochemistry. In: Magoon LB, Dow WG (eds) The petroleum system—from source to trap. AAPG Memoir. vol 60, pp 93–120

Peters KE, Hackley PC, Thomas JJ, Pomerantz AE (2018) Suppression of vitrinite reflectance by bitumen generated from liptinite during hydrous pyrolysis of artificial source rock. Org Geochem 125:220–228

Rentel K (1987) The combined maceral-microlithotype analysis for the characterization of reactive inertinites. Int J Coal Geol 9:77–86

Romero-Sarmiento MF, Rouzaud JN, Barnard S, Deldicque D, Thomas M, Littke R (2014) Evolution of Barnett shale organic carbon structure and nanostructure with increasing maturation. Org Geochem 71:7–16

Singh AK, Singh MP, Sharma M, Srivastava SK (2007) Microstructures and microtextures of natural cokes: a case study of heat-altered coking coals from the Jharia Coalfield, India. Int J Coal Geol 71:153–175

Singh AK, Singh MP, Sharma M (2008) Genesis of natural cokes: Some Indian examples. Int J Coal Geol 75:40–48

Stach E (1952) Die Vitrinit–Durit Mischungen in der petrographischen Kohlenanalyse. BrennstChem 33:368

Taylor GH, Teichmüller M, Davis A, Diessel CFK, Littke R, Robert P (1998) Organic petrology. Gebrüder Borntraeger, Berlin

Teichmüller M (1987) Recent advances in coalification studies and their application to geology. In: Scott AC (ed) Coal and coal-bearing strata: recent advances. Geol Soc London Spec Publ 32:127–169

Tissot BP, Welte DH (1978) Petroleum formation and occurrence; a new approach to oil and gas exploration. Springer-Verlag, Berlin, Heidelberg, New York

Tissot BP, Pelet R, Ungerer P (1987) Thermal history of sedimentary basins, maturation indices, and kinetics of oil and gas generation. Am Assoc Petrol Geol Bull 71:1445–1466

van Krevelen DW (1961) Coal: typology—chemistry—physics—constitution, 1st edn. Elsevier, Amsterdam, p 514

van Krevelen DW (1993) Coal: typology—chemistry—physics—constitution, 3rd edn. Elsevier, The Netherlands, p 979

Varma AK (1996) Influence of petrographical composition on coking behavior of inertinite rich coals. Int J Coal Geol 30:337–347

Varma AK (2002) Thermogravimetric investigations in prediction of coking behavior and coke properties derived from inertinite rich coals. Fuel 81:1321–1334

Welte DH (1965) Relation between petroleum and source rock. AAPG Bull 49:2249–2267

Wood DA, Hazra B (2017) Characterization of organic-rich shales for petroleum exploration & exploitation: a review—part 2: geochemistry, thermal maturity, isotopes and biomarkers. J Earth Sci 28(5):758–778

Chapter 3
Source-Rock Evaluation Using the Rock-Eval Technique

The Rock-Eval pyrolysis technique is a widely used and highly regarded tool used by petroleum geochemists for source-rock geochemical profiling. In its most common configuration, it is an open-system programmed-pyrolysis mechanism, whereby between predefined temperature thresholds ramped heating patterns are applied to carefully prepared samples. The samples are initially pyrolyzed in an inert atmosphere (generating petroleum fluids) and subsequently oxidized in an oxidizing environment. The key advantages of this method are: its ability to analyze samples swiftly, generating a suite of useful source-rock characterizing measurements in a liable and reproducible manner, and consumption of very small amounts of the sample during the analysis (making it attractive for wellbore core and cuttings analysis). Rock-Eval is now applied routinely to test wellbore and outcrop sequences (Espitalié et al. 1985, 1986; Peters and Cassa 1994; Sykes and Snowdon 2002). However, usage of only small samples for the analysis (5–60 mgs for different types of source-rocks), has led to questions about whether such low quantities of materials are actually representative of the source-rock characteristics of complex shale formations. It is therefore essential to run multiple samples of reference shales to demonstrate the reproducibility of the results obtained for specific formations.

The Rock-Eval device was introduced in 1977 by Institut Français du Pétrole (IFP) France (Espitalié et al. 1977). The different configurations of the Rock-Eval apparatus and their functioning are described by Espitalié et al. (1977) for Rock–Eval 1, Espitalié et al. (1985), and Espitalié and Bordenave (1993) for Rock–Eval 2 and 3. A significant upgrade of the Rock-Eval equipment occurred in the 1990s. The modification introduced then facilitated complete combustion and pyrolysis of different types of organic-matter, further increasing the reliability of the measured temperature-maxima of hydrocarbon cracking. The Rock-Eval 6 instrument that included these upgrades was introduced and commercialized in 1996 by Vinci Technologies and is currently still being sold. Some of the main differences between the earlier Rock-Eval 2 model and the current Rock-Eval 6 model are the final pyrolysis and oxidation temperatures. In the earlier version, the final pyrolysis temperatures were kept at 600 °C, while in the Rock-Eval 6 pyrolysis temperatures can be maintained up to 800 °C. Further, the final oxidation temperatures of 850 °C can also be achieved using the

© Springer Nature Switzerland AG 2019
B. Hazra et al., *Evaluation of Shale Source Rocks and Reservoirs*,
Petroleum Engineering, https://doi.org/10.1007/978-3-030-13042-8_3

Rock-Eval 6 device. This allows complete combustion of heavier refractory materials present in some shales. Further, in the earlier models of Rock-Eval, helium gas used as the carrier gas, while in Rock-Eval 6, nitrogen is used (Lafargue et al. 1998). Behar et al. (2001) examined the S2 values for a known standard sample at different weights (~5 to ~78 mgs), using helium and nitrogen as carrier gas. They observed, that when nitrogen was used as the carrier gas, approximately similar S2 values were observed at different weights of the sample ($\Delta S2 = 2$ mg HC/g rock). On the other hand, when helium was used as carrier gas, for the same sample the S2 values showed larger deviation ($\Delta S2 = 2$ mg HC/g rock). They also observed during kinetic studies that when nitrogen was used as carrier gas the S2 values were nearly similar at different heating rates. However, when helium was used as the carrier gas, the S2 values were observed to increase with increasing heating rates. These findings indicate the greater suitability of nitrogen as the carrier gas during Rock-Eval experiments. Although the concept of Rock-Eval was initially introduced for characterizing organic-rich petroleum source rocks, in recent years Rock-Eval has found widespread application for analyzing a wider range of unconventional hydrocarbon formations incorporating source rock and/or reservoir attributes, viz. shales, coals and other tight-formations (Romero-Sarmiento et al. 2016). Further, in recent years Rock-Eval has also found application for characterization of organic-matter in soils, and other near-surface, geologically recent lacustrine and marine sediments (Di Giovanni et al. 1998; Disnar et al. 2003; Sebag et al. 2006; Saenger et al. 2013).

3.1   Methodology and Different Parameters

The Rock-Eval technique, in one complete analysis-cycle, generates several important petroleum-generation-related metrics, such as the amount of free hydrocarbons present within the sample, the residual hydrocarbon content, the TOC content, the thermal maturity level of the sample, the amount of reactive organic-matter, the presence of carbonate minerals, and the quality/type of organic-matter present within the samples (Lafargue et al. 1998; Behar et al. 2001). The non-isothermal Rock-Eval analysis begins with an initial pyrolysis cycle whereby the samples are pyrolyzed in a nitrogen atmosphere from a starting temperature of 300 °C to final temperatures of 650 °C or 800 °C, depending on the target setting of the analyzer. The pyrolysis cycle is followed by an oxidation cycle, where the samples are combusted in presence of oxygen. There are different Rock-Eval built-in programs viz. 'basic/bulk-rock method' and 'pure organic matter method', which can be selected. That choice depends on the type of sample being analyzed (Vinci Technologies 2003).

In the basic/bulk-rock method, the sample is initially isothermally held at 300 °C for 3 min (Fig. 3.1). During this phase, the free hydrocarbon molecules and/or those that are easily vapourized or only loosely bound to the sample matrix are released forming the vapour that is recorded to constitute the S1 curve of Rock-Eval. Such released vapour is detected by the Flame Ionization Detector (FID). Following the 3-min initial isothermal step, the samples are heated during the pyrolysis cycle sys-

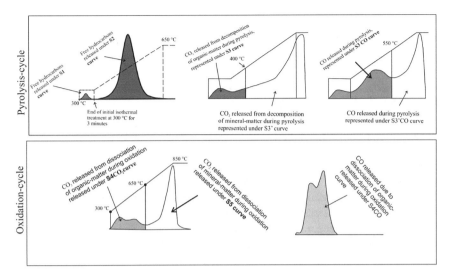

Fig. 3.1 Generalized diagram showing the S1–S5 Rock-Eval peaks typically derived using the basic Rock-Eval method (modified after Behar et al. 2001). See text for further discussions

tematically in small regular steps along a heating ramp from 300 °C to the set final temperature (650 or 800 °C). During this second-pyrolysis stage, the hydrocarbon molecules and structures within the organic-matter or kerogen are accessed, released and/or cracked to smaller more volatile hydrocarbon molecules. At these higher temperatures, akin to those used in cracking processes in crude oil refineries, the heavier-pyrolyzates within the kerogen are liberated and vapourized. These vapours are again detected by the FID and are recorded to constitute the S2 curve. Thus, the S2 values indicate the remaining petroleum-generating capability of the sample, i.e., the quantities of petroleum the sample could generate if it were taken naturally over geological time through the full thermal maturation cycle (into the dry gas zone of thermal maturity). The detailed shape and temperature characteristics of the S2 peak are extensively used in classifying the petroleum-generation potential of coal, shales and kerogens. The temperature maxima at which the maximum amount of pyrolyzates is generated within the S2 peak are referred to as T_{max}, which is widely used as a thermal maturity indicator (Fig. 3.1). The more thermally mature a sample, the more stable its residual hydrocarbon moieties (or molecular components) become, i.e., those left behind once the volatile or petroleum-like molecules are driven off as part of the S1 peak. The residual hydrocarbon moieties require higher temperatures to be broken down and generate further petroleum-related molecules.

A significant improvement in Rock-Eval 6 versus earlier models is the positioning of the probe monitoring pyrolysis temperatures. In the earlier Rock-Eval equipment, that temperature probe was not directly in contact with the sample-bearing crucible but was placed within the oven-wall. This meant that the temperatures recorded in the older Rock-Eval models were not actually the temperatures reached by the sample, but those reached by the oven-wall temperature. In the Rock-Eval 6 device, the thermo-couple is placed directly on the piston, thereby enabling a precise determination of temperature of the sample undergoing pyrolysis (Behar et al. 2001). For the earlier versions of Rock-Eval, the temperature-maximum of the S2 peak was called T_{max}, however in Rock-Eval 6, the actual temperature maximum recorded is the S2 temperature-peak. The T_{max} parameter in Rock-Eval 6 is thus an arbitrary maturity index applied on the S2 temperature-peak and does not represent the true temperature, and is only used to keep parity in the geochemical classification-scheme for analysts (Wood and Hazra 2018). In Rock-Eval 6, the S2 temperature-peak is the value obtained from the curve and is dependent on the temperature ramp used for specific analysis. The T_{max} metric, on the other hand, is calculated to yield approximately the same value for a specific sample at different heating-rates, and thereby provides a useful indicator of that samples thermal maturity. Hence, the temperature difference (ΔT) between S2 temperature-peak and T_{max} is slightly different when different Rock-Eval heating rates are used. Table 3.1 shows the results for the IFP160000 synthetic shale-standard analyzed at two different heating-rates of 25 and 5 °C/min. The Rock-Eval user's guideline suggests that the IFP160000 standard should have a T_{max} value of 416 ± 2 °C. At both the heating-rates, the T_{max} is observed to be close to that expected value, while the S2 temperature-peak is observed to be much higher at the faster heating-rate due to the laws of kinetics. Thus, the T_{max}, value derived for a sample by Rock-Eval should be similar or constant, irrespective of the heating-rates applied, and can therefore be used as a proxy for that samples level of thermal maturity. On the other hand, the S2 temperature-peak which represents the actual temperature should be used for kinetic analysis (i.e., to derive the activation energies of the reactions involved in generating the S2 peak).

During the pyrolysis of rocks containing organic-matter, the different oxygenated compounds present within the organic-matter are also decomposed, generating carbon monoxide (CO) and carbon dioxide (CO_2), which are detected by continuous sensitive on-line infrared detectors. The CO and CO_2 generated during pyrolysis can come from both organic and/or inorganic sources (particularly carbonate minerals). The CO_2 generated from decomposition of oxygenated groups of organic-matter, takes place between 300 and 400 °C, and this constitutes the S3 curve of Rock-Eval (Behar et al. 2001). The S3 curve is used for calculating oxygen indices [OI; OI = (S3/TOC) * 100] and TOC. The CO_2 from inorganic sources, is that derived from temperatures above 400 °C up to the end of pyrolysis temperature cycle, and is expressed as S3', which is used to determine the quantity of mineral carbon present in the sample. Similarly, the CO generated can also have organic and/or inorganic sources. These constitute the S3CO (CO from organic-matter) and S3'CO (CO from

Table 3.1 S2 temperature-peak and T_{max} of IFP160000 at different heating-rates

IFP160000 (°C/min)	T_{max} (°C)	S2 temperature-peak	ΔT
25	416	456	40
5	418	425	7

both organic and inorganic matter) Rock-Eval peaks (Fig. 3.1). The organic-CO component contributes to the overall Rock-Eval TOC calculation. Assuming, that the hydrocarbons released under S1 and S2 contain 83% carbon, the total pyrolyzable carbon (PC) is calculated (also incorporating the carbon released from oxygenated compounds) (Behar et al. 2001).

Following the Rock-Eval pyrolysis cycle, the crucible containing the pyrolyzed sample, is transferred robotically to an oxidation-chamber, where oxidation starts at 300 °C, burning off the remaining organic-matter, to a final temperature of 850 °C. The organic-matter when burnt in the presence of oxygen generates CO and CO_2, which are again detected by infra-red detectors and are expressed under Rock-Eval S4CO and S4 CO_2 curves (Fig. 3.1). The ramped-heating during the oxidation-cycle is maintained at 20 °C/min. The amount of CO and CO_2 generated from combustion of the organic-matter, contributes to the calculation of the residual carbon (RC) content of the sample. The combination of the PC and RC contents provides the quantified measure of TOC content for a sample. Typically, the entire organic-matter is combusted below 650 °C. If a sample contains carbonate minerals, such as calcite or dolomite, they decompose at higher oxidation temperatures (generally above 650 °C) and consequently the CO_2 generated due to oxidation of carbonate minerals constitute the S5 curve (Fig. 3.1). Other carbonate minerals such as siderite start decomposing during pyrolysis, at temperatures between 400 and 650 °C, which are represented under S3′ and S3′CO Rock-Eval curves, respectively (Lafargue et al. 1998).

The 'pure organic matter' Rock-Eval method is used for samples that do not contain any carbonate minerals. The final pyrolysis temperature in such cases is kept at 800 °C. This allows the determination of higher T_{max} values for decarbonated and mature coals. Further, during oxidation using the 'pure organic matter' method the entire CO_2 produced due to oxidation is represented under S4CO_2 curve, and the entire area under S4CO_2 curve is considered as RC for the purpose of TOC calculation. Since the samples are free from carbonates, there is no organic CO_2 (S4)-inorganic CO_2 (S5) boundary.

3.2 Important Indices and Critical Insights Regarding Rock-Eval Functioning for Source-Cum-Reservoir Rock Assessment

3.2.1 Rock-Eval S1

The Rock-Eval S1 peak indicates the presence or absence of free-to-move petroleum-related hydrocarbons (natural gases, oils and bitumens) present within a sample. As such it measures the gas and/or petroleum liquids/solids in situ in coal, shales and tight reservoirs. However, there is a chance that some of those free-to-move hydrocarbon components have actually migrated into the formation being tested (i.e. where generated in another formation but as part of the natural migration of petroleum have subsequently made their way into the sampled formation). While hydrocarbons released under S1 are most likely to contain an indigenous component, they can also consist of components migrated into the sampled formation or have been redistributed from other parts of that formation (Hunt 1996). Further, in wellbore samples the S1 peak may also contain a contribution from drilling-fluid contaminants; particularly if oil-based drilling fluids are employed. These hydrocarbon contaminants mainly crack during pyrolysis, generating false enhanced S1 peaks, but there residues may also affect the Rock-Eval S2 peak and its T_{max} values.

Carvajal-Ortiz and Gentzis (2015) documented the impacts on a Cretaceous shale the effects of oil-based-mud (OBM) contamination on the Rock-Eval S1, S2, and T_{max} measurements. For the initially contaminated sample, they observed extremely high S1 values, some 125 mv greater than the Rock-Eval 6 FID detection limits. Following sample cleaning using organic-solvents the sample was reanalyzed, revealing that the S2 and TOC in that contaminated samples were inflated by 36 and 19%, respectively, while the T_{max} value was suppressed by 16 °C compared to uncontaminated samples from the same formation. Cleaning contaminated samples with organic-solvents is may not be favorable always, as it will also remove part of the original S1 components and some of S2 components from the samples, along with the contaminants. Extreme care needs to be taken with wellbore samples to identify potential contamination before interpreting the data.

Figure 3.2, shows a typical pyrogram (S1 and S2 peaks only) for a sample significantly contaminated with OBM possessing a disproportionately large S1 peak (Fig. 3.2a), a pyrogram for an uncontaminated Permian shale from India (Fig. 3.2b), and a Permian shale pyrogram showing the presence of two smaller sub-peaks in-between S1 and S2 peaks. For the OBM contaminated shale (modified after Carvajal-Ortiz and Gentzis 2015), the hydrocarbons released under S1 is much larger than the hydrocarbons released under S2 (Fig. 3.2a). For the uncontaminated sample (Fig. 3.2b) the S1 peak is much smaller than S2 peak. It is also possible, in some samples, that there are slightly heavier hydrocarbon-phases and residues, which are not vapourized in the temperature range of the S1 peak but are vaporized either before or in the lower-temperature range of the S2-peak. These may then form sub-peaks between the main S1 and S2 peaks and can impact the petroleum-generation potential interpreted for that sample (e.g., Figure 3.2c). Dembicki (2017) opined that such sub-peaks can occur when a sample contains resins and/or asphaltenes which are not completely cracked by the S1 peak temperature range. Further, some drilling fluids contain Gilsonite (a form of asphaltene), that will release hydrocarbons in-between the S1 and S2 peaks in drilling-fluid-contaminated samples. As seen in Fig. 3.2c, the hydrocarbons released under the two sub-peaks are smaller than the main S2-peak, and the T_{max} value is not suppressed in this case. However, in highly contaminated cases it is possible that the peak representing the contaminants outgrow the true S2-peak for the sample, in which case there is a risk that the interpreted T_{max} value would be erroneously suppressed.

Hunt (1996) used a S1 versus TOC plot to distinguish non-indigenous and indigenous hydrocarbons present within a sample. S1/TOC values greater than 1.5 indicate the presence of migrated hydrocarbons or contaminants within the sample,

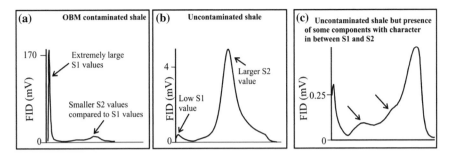

Fig. 3.2 Plots showing the relationship between hydrocarbons released under S1 and S2 curves of Rock-Eval. **a** shows the plot for an oil-based-mud(OBM) contaminated shale (modified after Carvajal-Ortiz and Gentzis 2015). The S1-peak disproportionately outgrows the S2-peak, yielding a false signature about the sample. **b** represents uncontaminated shale from India, where the S1 can be observed to be much smaller than the S2. **c** represents Lower Permian Barakar Formation shale from India, where two smaller sub-peaks (indicated by arrows) exists in-between S1 and S2 peaks. These are likely to be due to resins and/or asphaltenes in the sample (either indigenous or introduced as contaminants)

while values less than 1.5 indicate presence of indigenous or in situ hydrocarbons. In qualifying that threshold Jarvie (2012) suggested that the oil saturation index [OSI = (S1/TOC) * 100] should be greater than 100 mg HC/g TOC for saturated 'oil-producing' shales. However, OSI > 150 mg HC/g TOC OSI [equivalent to S1/TOC > 1.5], indicates the presence of drilling-contaminants or migrated oil in the sample.

Figure 3.3 shows S1 versus TOC for samples of Indian Permian shales from three coal basins viz. Raniganj, Jharia, and Auranga basin (Hazra et al. 2015; Mani et al. 2015; Varma et al. 2018), respectively. These samples all display S1/TOC ratios of <1.5 (OSI <150 mg HC/g TOC), indicating the presence of mostly indigenous hydrocarbons. The maximum OSI value for these samples is only 21.56 mg HC/g TOC, suggesting that they contain very low free gas/oil, and most of the petroleum present is in the S2 peak range (i.e., bound more comprehensively to the rock matrix or within the kerogen than the S1 component).

Figure 3.4 shows the S1 versus TOC cross-plot for Permian Lucagou Formation shales (China) from two basins viz. Junggar basin (Pan et al. 2016) and Santanghu basin (Zhang et al. 2018). Only two samples from the entire dataset display S1/TOC >1.5 and OSI >100 mg HC/g TOC. The occurrence of higher S1/TOC values for just those two samples suggests that contamination of the S1 content by drilling mud components may have occurred in those specific samples. An inspection of the Rock-Eval S1 and S2 peak curves can sometimes help interpreters to decide whether non-indigenous hydrocarbons are distorting the S1 peak values.

An extension to the established standard Rock-Eval heating ramp methodology is an approach called 'Shale Play' developed by IFP, with the specific focus of assessing shale-resource systems (Pillot et al. 2014a; Romero-Sarmiento et al. 2016). This is focused on providing more detailed analysis of the components of the S1 peak

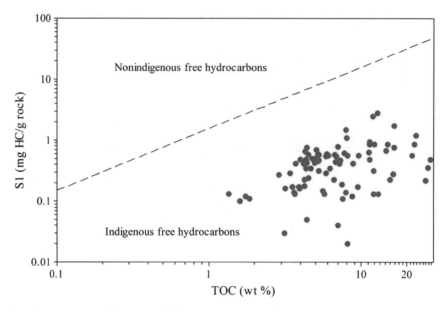

Fig. 3.3 Cross-plot of S1 peak and TOC content of samples of Permian shales from three Gondwana basins viz. Raniganj, Jharia, and Auranga, India (see text for sources)

and more easily discriminating non-indigenous hydrocarbons, in particular OBM contamination. The pyrolysis heating program starts at 100 °C and at a heating rate of 25 °C/min until the temperature reaches 200 °C. At 200 °C, the temperature is held constant for 3 min (i.e. a temporary temperature plateau). The hydrocarbons released during this phase are detected by the FID and are recorded and assigned to a Sh0 peak, composed of the lightest most-easily vaporized hydrocarbons in the sample, from 200 °C to the end of the temporary temperature plateau. The temperature is then raised to 350 °C, at a rate of 25 °C/min, and the temperature is again kept isothermal for 3 min (another temporary temperature plateau). The hydrocarbons released during this phase are assigned to the Sh1 peak and represents the heavier less-easily vaporized higher-molecular-weight hydrocarbons.

From 350 °C, the temperature is raised to 650 °C at a rate of 25 °C/min i.e., the standard S2 peak heating ramp) to generate the remaining pyrolyzable hydrocarbons constituting the S2 peak. In the 'Shale-Play' method the free hydrocarbons in a sample are denoted by the HC Content Index (HCcont) which is made up of Sh0 plus Sh1 peak components (mg HC/g rock) providing more detail for characterization than the traditional S1 peak provided by the bulk/basic Rock-Eval method. Based on experiments conducted using this method, Romero-Sarmiento et al. (2016) further suggested that more consistent T_{max} values could be obtained from the S2 peak for contaminated samples, without actually needing to clean those samples using organic solvents. These results suggest that the 'Shale Play' sample heating regime

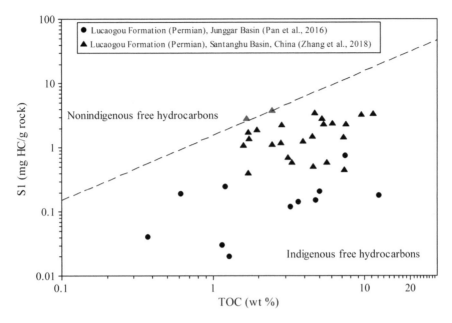

Fig. 3.4 Cross-plot of S1 peak and TOC content of Permian Lucagou Formation shales, China, from two basins viz. Junggar and Santanghu. Note that while for the majority of the samples, the hydrocarbons released under the S1 curve of Rock-Eval fall well within the designated indigenous range, for two samples the S1 values lay close to an S1/TOC ratio of 1.5, indicating the possible presence of migrated hydrocarbons or sample contamination by OBM during drilling

up to 350 °C more effectively segregates all the free hydrocarbons from a sample and assigns them to the Sh0 and Sh1 peaks.

3.2.2 Rock-Eval S2 Peak, FID Signals, HI, and T_{max}

The hydrocarbons released under S2 heating ramp range of temperatures (300–650 °C) during Rock-Eval provides insight to the remaining petroleum generation potential of an organic-rich rock sample S2 peak values can also help to identify the type of organic-matter present within the rock. When divided by the TOC, S2 values provide the hydrogen indices (HI; mg HC/g TOC) that can discriminate between different kerogen types present in a shale or coal (Lafargue et al. 1998). Type I and type II kerogens have the highest potential petroleum yield. Correspondingly, the highest S2 values and highest HI values are associated with type I- and type II-kerogen-bearing rocks. Type III-kerogen bearing rocks have much lower S2 peak and HI values and type IV-kerogen-bearing rocks have very small S2 peaks and very low HI values.

The Rock-Eval FID detector records the hydrocarbons released across the S1 and S2 temperature ranges. Rock-Eval 6 operator guidelines (Vinci Technologies 2003) suggest that the minimum and maximum detection limits for the FID device are 0.1 and 125 mV signals, respectively. The first-step of conducting any Rock-Eval analysis is to check whether the FID detector response is linear or not. Ideally the IFP standard (IFP160000) provided along with the device should be used to check the FID response, and the different parameters associated with its measurements. One way to cross-check whether the FID response is linear is by analyzing the IFP standard (with known values of different parameters) at different weights. Ideally, when the standard is analyzed using a range of sample weights, there should be an increase in FID signal with increasing sample weight. The Rock-Eval 6 operator guidelines (Vinci Technologies 2003) suggest using sample weights between 50 and 70 mgs for bulk rocks. The IFP160000 standard mimics type II shale, and at 25 °C heating rate, the S2 values of IFP 160000 standard should be 12.43 ± 0.50 mg HC/g rock. If substantially different S2 values are obtained for that standard a calibration issue exists for that specific Rock-Eval machine.

Figure 3.5 shows the results for the IFP160000 standard analyzed at 4 different weights of 54.10, 57.41, 58.52, and 69.47 mgs conducted to verify that a machine is suitably calibrated. Ideally, with increasing sample weight, the FID signal should increase, as greater amounts of hydrocarbons are liberated and detected by the FID.

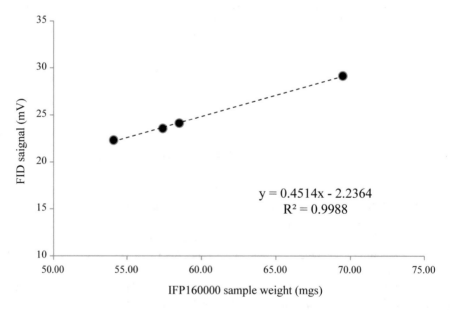

$$y = 0.4514x - 2.2364$$
$$R^2 = 0.9988$$

Fig. 3.5 Cross-plot showing the response of FID signals (mV) with increasing sample weights (mgs) tested using the IFP160000 standard with known values of the different Rock-Eval derived parameters. The strong positive linear correlation verifies that the FID is meaningfully calibrated. Such testing should be performed as an equipment verification test before analysts proceed to perform source-rock analysis on unknown samples

Also, when the data is converted to per gram rock/sample, the data should be more or less similar. For all the four IFP16000 weight-splits displayed in Fig. 3.5, the hydrocarbons liberated across the S2 temperature range were observed to vary between 12.00 and 12.68 mg HC/g rock, thus being within the guideline-specified permissible limits. Correspondingly, a strong correlation ($R^2 = 0.998$) is observed between the FID signal and sample charge for the Fig. 3.5 example. Such high correlation indicates the FID linearity (Fig. 3.5 and Table 3.2). Only after verifying the FID linearity, should analysts proceed to conduct Rock-Eval sample analysis.

HI values of immature type I, II, III, and IV kerogen are typically >600, 600–300, 200–50, and <50 mg HC/g TOC, respectively (Peters and Cassa 1994). HI values in between 200 and 300 mg HC/g TOC can be indicative of the presence of a mixture of type II and type III kerogens that is observed in many shales. However, predicting kerogen-type based on only HI can be erroneous and flawed (Behar and Vandenbroucke 1987; Hazra et al. 2015). With increasing thermal maturity levels, as hydrocarbons are liberated and expelled, the HI decreases, and correspondingly for a suite of samples across a wide-range of maturity, a negative relationship may be observed between HI and thermal maturity indices like T_{max} and Ro. For example, Hazra et al. (2015) for a suite of samples from the Raniganj basin India belonging to the Early Permian Barakar Formation showed that for samples at oil-generation-window thermal maturity levels, the HI values didn't show any relationship with T_{max} values (Fig. 3.6). However, for samples with higher maturities beyond the oil-window they observed a strong negative relationship between HI and thermal maturity. Frequently, suites of immature shales display variable HI depending upon the type of kerogen present, but with increasing maturity HI decreases and such distinctions become obscured (Behar and Vandenbroucke 1987; Behar et al. 1992).

Figure 3.6 shows the HI versus T_{max} plot for sixty-three Permian shales from Raniganj basin (Hazra et al. 2015) and Jharia basin (Mani et al. 2015), India. It can be seen that in general there is a decreasing HI trend of the shales with increasing T_{max}, especially beyond the oil-window. If HI of such samples is used to predict the type of organic-matter or kerogen present, the interpretation may be flawed. For example, for the six shales from Raniganj basin with T_{max} values >450 °C, the HI values vary between 40 and 101 mg HC/g TOC i.e. indicating presence of type III and type IV kerogen. However, petrographic results (Hazra et al. 2015) indicate the presence of type II kerogen, in addition to type III and type IV kerogen within the samples. Table 3.3 shows the results of those shales with T_{max} values greater than >450 °C.

Table 3.2 FID signals and S2 values for the IFP160000 standard at different weights	Sample weight (mgs)	S2 (mg HC/g rock)	FID signal (mV)
	54.10	12.68	22.310
	57.41	12.19	23.561
	58.52	12.00	24.127
	69.47	12.01	29.160

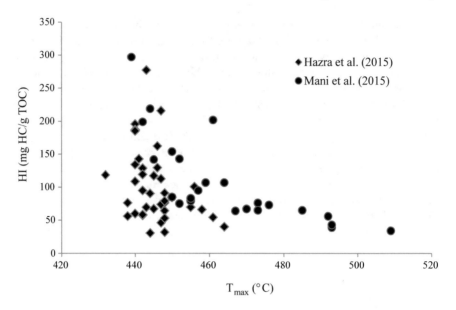

Fig. 3.6 Cross-plot showing the relationship between HI and T_{max} for shales from Raniganj basin (Hazra et al. 2015) and Jharia basin (Mani et al. 2015), India

Table 3.3 HI, T_{max}, and organic-matter types for shales from Raniganj basin with T_{max} values >450 °C

S.N.	HI (mg HC/g TOC)	T_{max} (°C)	V^{mmf} (vol. %)	I^{mmf} (vol. %)	L^{mmf} (vol. %)	A^{mmf} (vol. %)
CG 1263	85	450	58.64	41.36	0.00	0.00
CG 1001	70	455	49.62	30.83	15.37	4.18
CG 1283	66	458	57.69	36.37	5.95	0.00
CG 1284	54	461	49.38	41.36	9.27	0.00
CG 1285	40	464	68.22	29.62	2.16	0.00
CG 1286	101	456	59.70	33.83	6.47	0.00

Source Hazra et al. 2015
Note V^{mmf}, I^{mmf}, L^{mmf}, and A^{mmf} denote volume percentage of vitrinites, inertinites, liptinites (other than alginites), and alginites on mineral matter free basis, respectively

Samples CG 1283-CG 1286 (Lower Permian, Barakar Formation) contain some type II liptinites (other than alginites) which is not reflected by their HI values. Further, for sample CG 1001 (Upper Permian, Barren Measures Formation) in addition to type II, type III, and type IV organic-matter, alginites (type I) were also observed. These findings suggest that predicting the type of kerogen using Rock-Eval-generated data and present-day HI can be erroneous. Few attempts have been made to predict original HI of the organic-matter, and compare that with the Rock-Eval derived present-day HI to understand how much kerogen transformation has been completed.

The HI values of any suite of sample are dependent on the thermal maturity levels of the component organic-matter. However, the impact of the type of organic-matter present can also be critical on the thermal-maturity levels and concomitant transformation of organic-matter. For example, for a suite of early mature to peak mature (T_{max} < 450 °C) shale samples belonging to the Upper Permian, Barren Measures Formation from Raniganj basin, Hazra et al. (2015) observed lower-HI (67–277 mg HC/g TOC), although petrographic examination revealed the presence of alginites (type I kerogen) within them (Table 3.4). Type I kerogens have the highest HI values (>600 mg HC/g TOC). This indicates that although the samples are within the early stages of maturity, significant hydrocarbons have been generated with concomitant transformation of kerogen due to the presence of alginites which cracks at lower thermal maturity levels. Thus HI values at lower thermal maturity levels can also be misleading due to different responses of different types of organic-matter with thermal maturities. It is thus recommended that Rock-Eval HI (which indicates the present-day residual potential) should always be cross-checked with independently-derived information, such as organic petrography, for better approximation of the organic-matter type present within the rocks.

A closer inspection of the Rock-Eval S2 pyrograms usually helps to predict the different parameters. Figure 3.7 shows the Rock-Eval S2 pyrogram for the IFP160000 (synthetic) shale standard. It represents an immature shale (T_{max}: 416 ± 2 °C), with HI value (~380 mg HC/g TOC) indicating the presence of type II organic-matter. At 69.47 mgs sample weight, the FID signal, S2, and HI are 29.16 mV, 12.01 mg HC/g rock, and 376 mg HC/g TOC, respectively. A tighter Gaussian shape of the S2 pyrogram peak is associated with this type II kerogen.

Figure 3.8 shows the S2 pyrogram for a type I kerogen bearing Norwegian Geo-chemical Standard (shale), JR-1. For the JR-1 sample, the S2 pyrogram peak displays

Table 3.4 HI, T_{max}, and organic-matter types for shales from Raniganj basin with T_{max} values <450 °C

S.N.	HI (mg HC/g TOC)	T_{max} (°C)	V^{mmf} (vol.%)	I^{mmf} (vol.%)	L^{mmf} (vol.%)	A^{mmf} (vol.%)
CG 1002	91	448	56.86	26.16	12.66	4.32
CG 1003	67	445	55.93	31.52	10.27	38
CG 1004	95	442	48.48	37.23	12.55	1.73
CG 1005	134	440	44.69	26.33	12.83	16.15
CG 1006	129	442	48.62	24.07	14.85	12.46
CG 1007	74	447	57.13	31.60	7.77	3.50
CG 1008	79	448	50.27	41.25	7.64	0.84
CG 1009	119	442	48.94	25.07	10.34	15.65
CG 1010	185	440	534	17.91	12.44	17.41
CG 1011	277	443	44.22	18.09	14.11	23.58
CG 1012	119	432	51.01	21.89	16.20	10.90

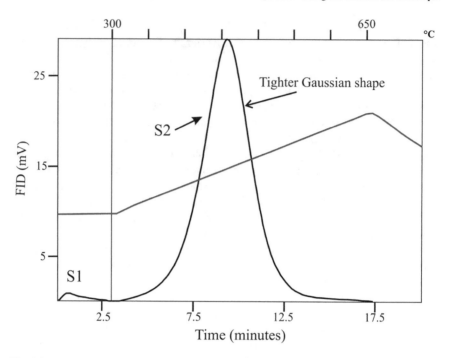

Fig. 3.7 Rock-Eval S2 pyrogram of IFP160000 shale standard at 69.47 mgs weight. The FID signal was observed to be 29.16 mV. Note the tighter Gaussian shape of the S2 pyrogram peak, which contrasts with the S2 pyrogram peak shape for type III–IV organic-matter bearing rocks (see Fig. 3.10). The red line represents the FID temperature line at a ramp of 25 °C/min

an even tighter Gaussian shape. At 10.48 mgs weight, the FID signal for the JR-1 is >35 mV, indicative of a high oil-generating kerogen within the sample. Clues such as this, along with thorough inspection of the S2 pyrograms can help the interpreters to infer the type of organic-matter present. At 5.7 mgs weight, the FID signal for JR-1 is >17 mV. Considering FID linearity with sample size, for a JR-1 sample weight of about 35 mgs or more, there is a risk of the S2 peak saturating the FID detector.

Comparing the FID signals for kerogens of type II (mimicking IF160000 standard) and type I (mimicking JR-1 standard), it is evident that FID signals for type I kerogen are higher. Consequently, the chances for FID saturation is higher for type I kerogen-bearing rocks at higher sample weights. The problems of FID saturation while analyzing type I shales using Rock-Eval 6, and the revised methodology needed for reliable data generation is provided by Carvajal-Ortiz and Gentzis (2015). They demonstrated that for a Green River shale sample from Utah (U.S.A), at approximately 60 mgs of sample weight, the FID signal was close to 600 mV, well beyond the FID saturation limits and thereby resulting in a broader S2 pyrogram peak. When they reanalyzed the samples at lower weights (of approximately 5 mg), the S2 pyrogram peak displayed a tighter Gaussian shape, associated with a reliable FID signal below the saturation limit. Further, by lowering the sample weights for the Green

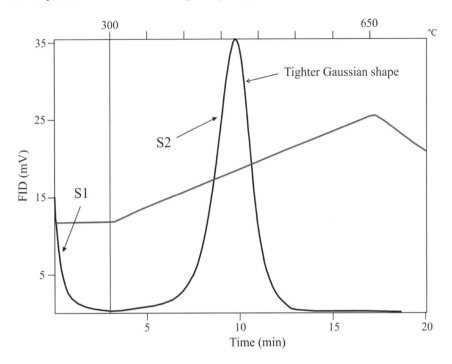

Fig. 3.8 Rock-Eval S2 pyrogram of Norwegian Geochemical Standard shale JR-1 at 10.48 mgs weight. The FID signal was observed to be >35 mV. Note the tighter Gaussian shape of the S2 pyrogram peak, which is contrast to the S2 pyrogram peak shape for type III-IV organic-matter bearing rocks (see Fig. 3.10). The S2 pyrogram peak for type I kerogen-bearing JR-1 is tighter in shape than type II kerogen-mimicking IFP160000 standard. The red line represents the FID temperature line at a ramp of 25 °C/min

River shale, Carvajal-Ortiz and Gentzis (2015) also observed that the TOC, S2, and T_{max} calculations became more precise. This example demonstrates that monitoring the FID signals and the shape of the S2 pyrogram peaks, and not just the Rock-Eval derived data, is extremely important for reliable source-rock interpretations. For type I shales, such as the JR-1 standard and the Green River shale, the FID signals should be closely monitored by the interpreters. It is considered appropriate for generating reliable data for high oil-yielding shales/bulk rocks (i.e., high HI materials), the sample weights analyzed should ideally be smaller; ideally below 35 mg rather than the more typical of 50–70 mgs.

In contrast to FID detector saturation associated with some type I and type II kerogen samples, in some cases, due to the presence of minimal or overmature or low hydrocarbon-yielding organic-matter, the FID signals may be too low. This outcome also results in misleading results. The Rock-Eval user's manual mentions that the minimum FID signal for detection should be 0.1 mV (Vinci Technologies 2003).

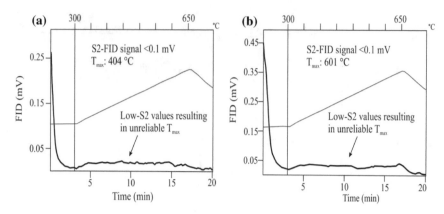

Fig. 3.9 Rock-Eval S2 pyrograms of low hydrocarbon-yielding, low-TOC bearing Proterozoic shales from Vindhyan basin, India. The S2 FID signals for both the samples are lower than the minimum reliable Rock-Eval 6 FID signal of 0.1 mV. Consequently, the data produced for both the samples are not reliable. For sample shown in (**a**), T_{max} value of 404 °C was derived from S2 peak analysis, while for sample shown in (**b**), T_{max} value of 601 °C was similarly derived. The red lines represent the FID temperature line at a ramp of 25 °C/min

Any signal lower than this value, due to the type of kerogen material present, may not represent the sample properly, and thereby creating possibilities of erroneous interpretations.

Figure 3.9 shows S2 pyrogram peaks for two shale samples from the Proterozoic Vindhyan basin, India, studied earlier by Dayal et al. (2014). While both the samples are from the same suite, one generated a lower T_{max} value of 404 °C (Fig. 3.9a), while the other showed extremely high T_{max} value of 601 °C (Fig. 3.9b). These discrepant results for the two samples from the same suite [collected from outcrops (Dayal et al. 2014)], are due to the generation of minimal quantities of hydrocarbons that are below the minimum reliable FID detection limits, and not related to the samples being immature or overmature. Under such circumstances, the software imprecisely detects some peak point across a broad and poorly defined S2 peak, and thus the T_{max} value for such samples is misleading in relation to their petroleum-generation potential. For the samples shown in Fig. 3.9a, b the S2 values were observed to be 0.02 and 0.03 mg HC/g rock, respectively. These examples highlight the need to check for anomalies or out-of-recommended range FID counts in pyrograms before drawing conclusions from the pyrogram data regarding petroleum-generation potential of the samples.

For type III-IV kerogen-bearing shales, the Rock-Eval pyrograms also show distinguishing S2-peak characteristics that typically includes an exaggerated right-side tail, i.e., an extended decaying limb towards the end of the pyrolysis S2 heating cycle. While the S2 curves can be tighter (Fig. 3.10a) or open (Fig. 3.10b) in nature, the tailed-effect is observed mostly in all type III-IV kerogen bearing samples. For both the samples shown in Fig. 3.10, it is apparent that the entire petroleum generated during the S2 heating cycle is not released within the upper-temperature limit of

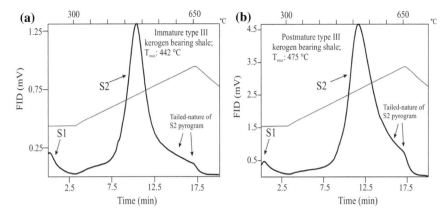

Fig. 3.10 Rock-Eval S2 pyrograms of two Lower Permian Barakar Formation shale samples from the Jharia basin, India. For both the samples, the Rock-Eval S2 FID signals are of a much lower magnitude than the JR-1 and IFP16000 standards. For sample A, the T_{max}, HI, and TOC values are 442 °C, 88 mg HC/g and 2.97 wt%, respectively. For sample B these metrics display values of 475 °C, 78 mg HC/g and 22.78 wt%, respectively. For both the samples the tailed-effect on the decaying limb (right shoulder) of the S2 pyrogram peak (identified by arrows), is characteristic of type III-IV kerogen-bearing samples. Petrographic studies show presence of vitrinites, inertinites, and liptinites (other than alginites) in the proportions of 54, 44, and 2 vol.% (mmf basis) in sample A, and 54.5, 40.8, and 4.7 vol. (mmf basis) in sample B, respectively. The red lines represent the FID temperature line at a ramp of 25 °C/min

650 °C of that pyrolysis-cycle. Typically, the S2 (right-side) decaying limb reaches the base of the pyrogram FID scale due to the truncation of the heating cycle.

One of the main reasons for increasing the upper-temperature pyrolysis limits for the Rock-Eval 6 device to 850 °C was to facilitate complete thermal decay of type III kerogen during the S2 heating ramp (Lafargue et al. 1998). A pyrogram S2 curve with a right-sided tail can be used as an indirect indicator of type III kerogen in a sample. Figure 3.10 shows S2 pyrograms of two Permian shale samples from the Jharia basin, India, containing type III kerogen. The difference in the shapes of type I, II and III kerogens (Figs. 3.7, 3.8 and 3.10), are caused, at least in part, by their distinct chemical structures. Liptinite macerals are marked by long-chained and less-branched aliphatic components, making them more thermally labile. In general, liptinites are characterized by the longest aliphatic chains and highest hydrocarbon generation capability among the three maceral groups. In contrast, inertinite shows the least hydrocarbon generation capacity and highest aromaticity. Vitrinites tend to display characteristics that are intermediate between liptinites and inertinite (Chen et al. 2012). Among the liptinite macerals, there is some variation in their relative aliphatic contents. Generally, aliginite macerals (type I) are marked by more abundant aliphatic structures and less abundant aromatic structures, and is followed by the resinite, cutinite and sporinite macerals (type II) i.e. decreasing aliphatic to aromatic ratio (Guo and Bustin 1998). This labile nature enables type I and II kerogens to crack at lower pyrolysis-temperatures. Type III-IV kerogen on the other hand is marked by the presence of more abundant aromatic structures and less-abundant aliphatic structures. The lower hydrogen contents of type III kerogens makes them thermally less labile, and thereby requiring higher pyrolysis temperatures for complete thermal degradations. Comparison of the S2 pyrogram peaks for the two samples shown in Fig. 3.10 also illustrates the influence of thermal maturity on hydrocarbon generation from type III-IV kerogen-bearing samples. Owing to the early-mature nature for sample A, the right-side tail of the S2 peak starts at a lower temperature (Fig. 3.10a), while for sample B the right-side tail becomes evident at a higher pyrolysis temperature due to its thermally postmature state (Fig. 3.10b).

Impact of sample crush-size on Rock-Eval S2 peak characteristics

The user's guide for the Rock-Eval 6 instrument mentions that particle crush-size for a sample can be within the range of 2 mm to 100 μm. The impact of sample crush-size on Rock-Eval S2 peak shape and other parameters can be significant and influence the source-rock characterization conclusions drawn. Jüntgen (1984) identified that sample crush-size affects the different reaction pathways during pyrolysis experiments. Wagner et al. (1985) proposed that with increasing crush-size, the liberation and escape of volatiles from the particles becomes restricted, and the associated chemical-reaction rates become less significant. Inan et al. (1998) demonstrated the impact of particle crush-size on Rock-Eval parameters such as T_{max} and S1, S2, and S3 temperatures. Further, Hazra et al. (2017) while analyzing a vitrain band (manually-isolated from a coal sample) observed increased Rock-Eval S1 and S2 by lowering of the particle crush-size of the samples analyzed.

Table 3.5 shows the results for three shale samples with varying TOC contents for two contrasting particle crush-sizes of 1 mm and <212 microns. For the organic-rich shales, increase in S2 and S1 peak magnitudes with decreasing particle crush-size was observed to be more pronounced than for the organic-lean low-TOC shale. Figure 3.11 shows the Rock-Eval S1 and S2 curves for a high-TOC shale sample from the Raniganj Formation (Raniganj basin, India). By lowering the particle crush-size from 1 mm to 212 microns, the FID counts recorded for both S1 and S2 peaks increase, and thereby yield higher S1 and S2 peak magnitudes and TOC values. T_{max} on the other hand was observed to be almost identical for both particle-crush sizes. The quantum of increase in S1 and S2 peak magnitude and TOC, with decreasing particle-crush sizes, was found to be more pronounced for a carbonaceous shale sample (belonging to the Barakar Formation from Birbhum basin, India) compared to the high-TOC shale (Fig. 3.12). On the other hand, similar S1 and S2 peak magnitude and TOC values at both particle crush-sizes were observed for a low-TOC Intertrappean shale sample from Birbhum basin, India (Fig. 3.13).

The impact of crush-size on Rock-Eval parameters seems to be potentially more significant for organic-rich sediments, compared to organic-lean sediments. Organic-rich sediments, owing to their higher volatile contents, on being crushed to finer sizes, allow the volatiles to escape more easily. This, in turn seems to facilitate more thermal decomposition (chemical reaction) of the organic-matter. On the other hand, for organic-lean sediments, owing to their inherently lower volatile contents and higher mineral matter contents, the impact of a reduced sample size has minimal impact on the thermal decomposition of their more dispersed organic-matter.

Table 3.5 Rock-Eval results for three shale samples from India across two contrasting particle crush-sizes

Sample type	Basin/formation	Particle crush-size	S1 (mg HC/g rock)	S2 (mg HC/g rock)	T$_{max}$ (°C)	TOC (wt%)	HI (mg HC/g TOC)
Low-TOC shale	Birbhum/Intertrappean	1 mm	0.02	0.66	454	1.14	58
		<212 microns	0.03	0.66	456	1.14	58
High-TOC shale	Raniganj/Raniganj	1 mm	0.24	7.42	432	6.72	110
		<212 microns	0.34	10.68	433	9.08	118
Carbonaceous shale	Birbhum/Barakar	1 mm	0.34	19.74	434	19.22	103
		<212 microns	0.51	30.09	434	23.82	126

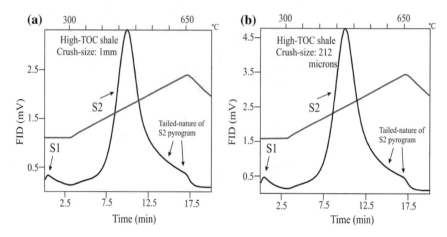

Fig. 3.11 Rock-Eval S2 pyrograms of a Upper Permian Raniganj Formation shale sample (with high-TOC content; 9.08 wt%) from the Raniganj basin, India, at two sample crush-sizes of 1 mm (**a**) and <212 microns (**b**). Note the similar S2 peak shapes but contrasting FID count scales. The red lines represent the FID temperature line at a ramp of 25 °C/min

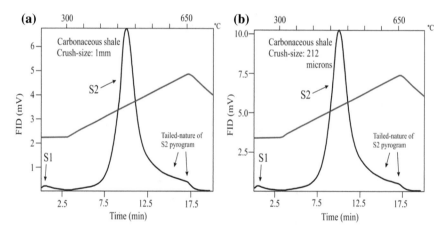

Fig. 3.12 Rock-Eval S2 pyrograms of a Lower Permian Barakar Formation carbonaceous shale sample (very high TOC content; 23.82 wt%) from the Birbhum basin, India, at two sample crush-sizes of 1 mm (**a**) and < 212 microns (**b**). Note the similar S2 peak shapes but contrasting FID count scales. The red lines represent the FID temperature line at a ramp of 25 °C/min

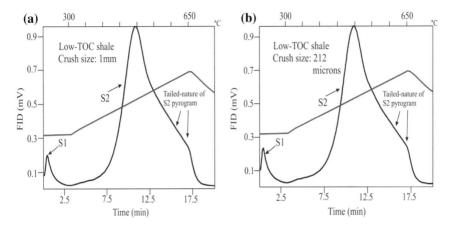

Fig. 3.13 Rock-Eval S2 pyrograms of a Lower Permian Barakar Formation shale sample (very low TOC content; 1.14 wt%) from the Birbhum basin, India, at two sample crush-sizes of 1 mm (**a**) and <212 microns (**b**). Note the similar S2 peak shapes and identical FID count scales. The red lines represent the FID temperature line at a ramp of 25 °C/min

3.2.3 Rock-Eval S3, OI

Similar to the hydrogen index, the oxygen index (OI) is a very important ratio derived from Rock-Eval analysis. The amount of CO_2 generated from organic-matter during the pyrolysis stage, is recorded by the S3 curve of Rock-Eval. When expressed as a ratio relative to the TOC it provides the OI [OI = (S3/TOC) * 100]. In terms of the programmed pyrolysis, the S3 peak represents the CO_2 released from organic-matter during the time of S1 release together with CO_2 released up to 400 °C during the S2 pyrolysis heating ramp. CO_2 recorded at temperatures above 400 °C is primarily derived from carbonate-bearing minerals and is thus designated as S3′, which is used to distinguish pyrolyzable mineral carbon. Espitalié et al. (1977) established this distinct origin of CO_2 released during pyrolysis using thermogravimetric analysis, identifying that no carbonate minerals decompose at temperatures below 400 °C during pyrolysis. Lafargue et al. (1998) established that the most common carbonate minerals such as calcite and dolomite decompose during oxidation at temperatures exceeding 650 °C. However, carbonate mineral such as siderite begin decomposing and generating CO_2, during pyrolysis stage at temperatures between 400 and 650 °C. Figure 3.14 shows the S3CO$_2$ and S3′CO$_2$ pyrolysis graphic of the IFP160000 standard. The IFP synthetic standard shows the presence of pyrolyzable organic-CO_2 (coming from organic-matter) and pyrolyzable inorganic-CO_2 (coming from mineral matter). The Rock Eval manual (Vinci Technologies 2003) mentions the acceptable S3 value of the standard to be 0.79 ± 0.20 mg CO_2/g rock. For the IFP 160000 standard shown in Fig. 3.14, the S3 was observed in an example analysis to be 0.65 mg CO_2/g rock.

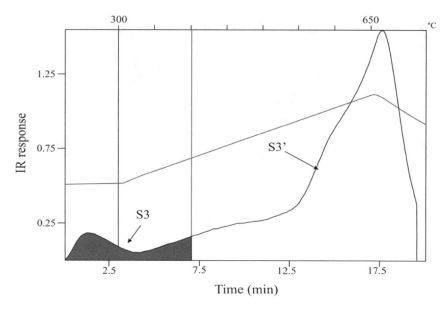

Fig. 3.14 S3 and S3′ pyrolysis (CO_2) graphic of IFP160000 synthetic shale-standard. For the calculation of OI, only the CO_2 released from the start of pyrolysis to 400 °C is considered (organic-CO_2). The CO_2 released above 400 °C is considered to have inorganic sources (coming from carbonate minerals) and should not be used for calculating OI or TOC. However, the presence of carbonate minerals is known to impact the magnitude of the S3 peak and associated OI values (see text for further discussions). The red line represents the FID temperature line at a ramp of 25 °C/min

The CO_2 and CO generated from organic-matter during pyrolysis are essentially derived from the oxygen-bearing functional groups present within the organic-matter (Lafargue et al. 1998). Similar to the atomic H/C versus O/C plot of van Krevelen (1961) for kerogen classification, Espitalié et al. (1977) used HI versus OI cross-plots to generate similar information. However, the OI can be impacted both by the presence of carbonate minerals within the host rock, and by the level of thermal maturity. Katz (1983) highlighted the tremendous potential impact of carbonate minerals on Rock-Eval's S3 peak and the OI. For samples from a formation made up of more than 70% calcite and dolomite and low-TOC content, Katz (1983) observed higher S3 and OI values. However, when the carbonate-minerals were removed from the tested samples by treatment with hydrochloric acid, and reanalyzed, a sharp reduction in the magnitude of the S3 peak and OI values resulted. Due to the generation of inorganic CO_2, formed by the dissociation of carbonate minerals in the untreated samples, a significant scatter of OI values was observed. This effect means that OI values should be treated with caution for carbonate bearing formation, before plotting the measured values on a van Krevelen diagram in attempts to assess the thermal maturity and/or kerogen-type in tested samples.

Typically, with increasing thermal maturity, the organic-matter in a formation becomes more enriched in carbon with progressive elimination of oxygenated com-

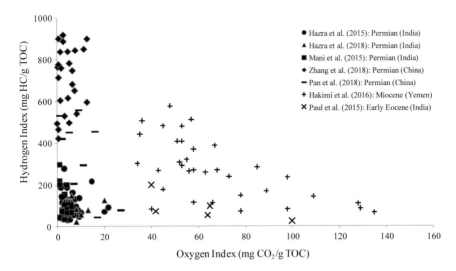

Fig. 3.15 Hydrogen index (HI) versus oxygen index (OI) cross-plot for shales from different countries and geological ages. Shales of more recent geologically origin tends to be marked by much higher OI values compared to geologically older samples

pounds and hydrogen-bearing compounds. Correspondingly, with increasing thermal maturity levels, in terms of Rock-Eval indices, both HI and OI values should reduce as thermal maturity increases. Figure 3.15 shows a cross-plot between HI and OI for shales across different countries and geological ages. It reveals that OI values of geologically recent sediments (Miocene and Early Eocene) have higher OI values, irrespective of their HI values. Moreover, the geologically older Permian samples from India and China display extremely low OI values irrespective of their HI values. For any kerogen-bearing formation its HI values primarily depend upon the type of kerogen present within them, and with increasing thermal maturity levels the HI values should decrease. Similarly, the OI values should also decrease with increasing thermal maturity levels. Permian shales from India (Hazra et al. 2015; Mani et al. 2015) typically display very low OI values (mostly <10 mg CO_2/g TOC; Fig. 3.14), with HI values ranging between 300 and 30 mg HC/g TOC (dominantly type III kerogen) and thermal maturity levels spanning a wide range of maturity from immature to overmature (Fig. 3.6). On the other hand, Permian shales from Junggar Basin China (Pan et al. 2016; Zhang et al. 2018) display very low OI values (mostly <10 mg CO_2/g TOC; Fig. 3.15) associated with higher HI values (mostly >300 mg HC/g TOC) than those recorded for the Indian Permian shales.

While OI can be used to demarcate geologically distinct samples (as seen for the geologically recent vs. old rocks), the very low OI values often associated with geologically older organic-rich formations with different organic-matter types restricts the use of OI as a maturity index. Notably lower values of OI and a lack of any relationship between OI and T_{max}, was also observed by Kotarba et al. (2002) in Upper Carboniferous Polish coals and shales, spanning a wide range of thermal maturity.

On the other hand, the same samples displayed negative relationship between atomic O/C ratios and thermal maturity levels. They opined that these anomalously low OI values could be due to the presence of relatively stable oxygen moieties which are not thermally-cracked at lower pyrolysis temperatures. Under such circumstances, a portion of the CO_2 that really belongs to the S3 peak may have its release delayed to higher temperatures and be inappropriately recorded as part of the S3′ peak. Such an eventuality would lower the calculation of TOC (marginally) and lead to erroneous source-rock interpretations. A critical evaluation, specifically targeting a range of coal and shale measures displaying lower OI values, and their relationship with carbonate-mineralogy, might reveal some fundamental properties associated with the cracking of stable oxygen moieties in relation to pyrolysis temperatures.

Other Rock-Eval derived parameters

- **Production Index (PI)**: The ratio of the hydrocarbons liberated under the S1 curve to the total amount of hydrocarbons released under S1 and S2 curves combined, gives the production index (PI) (Peters and Cassa 1994). Similar to Rock-Eval T_{max}, PI is used as a maturity proxy. However, similar to S1, higher PI values of rocks can indicate presence of migrated hydrocarbons or contamination from external sources (Hunt 1996). Peters (1986) observed that PI usually varies from 0.1 in immature sediments or at the start of oil-window, to 0.4 for mature sediments at the threshold of the higher end of the oil-generation window into the early stages of the wet-gas-generating window, to 1 for sediments whose hydrocarbon generation potential has been completed. To be confident in interpreting the PI consistently, the S1 and S2 signals, and the relationship between S1 and S2 should be monitored closely (as discussed in preceding sections). For example, the Proterozoic Vindhyan shales with extremely low FID signals and grossly different T_{max} values shown in Fig. 3.9a, showed PI of >0.60, providing a false impression of the host-rock's petroleum-generation potential. A cross-plot between PI and T_{max}, helps in elucidating the maturation, nature of the hydrocarbon products (migrated or in situ) of a source-rock (Hakimi et al. 2016).
- **Potential Yield (PY) or Genetic Potential (GP)**: The total amount of hydrocarbons generated/liberated under pyrolysis S1 and S2 curves is known as potential yield (Ghori 1998) or genetic potential (Varma et al. 2014, 2015). This index helps in categorizing the target horizon in terms of total petroleum-generation potential.

3.2.4 CO₂ Associated with S4 and S5 Peaks, and Relating to TOC and Mineral Carbon

The Rock-Eval S4 peak measurements tend to be less-widely used than S1, S2 and S3 peak data, despite containing information that assists in the determination of residual carbon (RC) and TOC of potential source-rocks. Lafargue et al. (1998) and Behar et al. (2001) identified the Rock-Eval S4 data and the value in monitoring the magnitude of the S4 peak was further highlighted by Hazra et al. (2017). If an organic-rich sedimentary rock contains carbonate minerals, the CO_2 curve displays a minimum detected response between 550 and 720 °C (generally ~650 °C) separating the S4 and S5 peaks. This minimum is related to the decomposition temperature-ranges of organic-matter and mineral matter. The CO_2 curve is divided into organic-CO_2 (S4CO_2; between 300 to the temperature associated with the IR response minima) and inorganic CO_2 (S5; from the temperature of the IR response minima to the final pyrolysis temperatures) (Behar et al. 2001).

Figure 3.16 displays S4CO_2 and S5 oxidation graphics for the IFP160000 and JR-1 standards, distinguishing the organic and inorganic CO_2 curves. The S4CO_2 and S5 curve characteristics associated with the two standards is typical for carbonate-bearing organic-rich formations. However, for certain formations there can be some doubt as to whether the CO_2 coming from organic-matter is completely liberated below the IR detection minima defining the boundary between the S4 and S5 peaks. For instance, some organic-rich shales/coals with higher fractions of RC content relative to the PC content can provide anomalous results based on the relative magnitudes of their S4 and S5 peaks. Type III-IV kerogen-bearing shales and coals from India (Hazra et al. 2017), tested for a range of sample sizes, and resulted in CO_2 eluting from the organic-matter over a wider temperature-range, in some cases, extending beyond the boundary between the S4 and S5 peaks. Consequently, a portion of the organic-CO_2 is incorrectly recorded under the S5 peak in some cases. This results in discrepant lower estimates of TOC and higher estimates of HI [as HI = (S2/TOC) * 100].

Figure 3.17 displays the S4CO_2 oxidation graphics of type III kerogen-bearing Lower Permian Barakar Formation carbonaceous shale, devoid of any carbonate

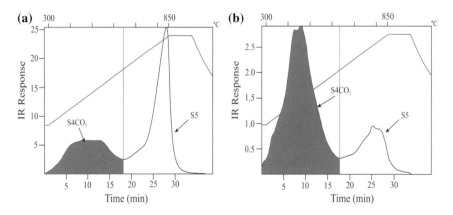

Fig. 3.16 S4CO_2 and S5 oxidation graphics from the Rock-Eval oxidation stage, for IFP160000 standard (**a**) and JR-1 standard (**b**). S4CO_2 represents the CO_2 generated from organic-sources between 300 °C to a minimum IR response that typically occurs between 550 and 720 °C (blue line at 650 °C). S5 represents the CO_2 generated from inorganic-sources from the minimum IR response that occurs between 550 and 720 °C (blue line) to the final experimental temperatures. The red lines represent the IR temperature line at a ramp of 20 °C/min

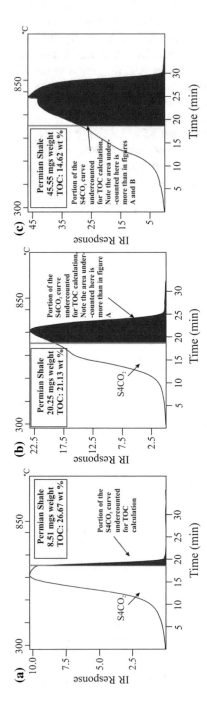

Fig. 3.17 S4CO₂ and S5 graphics from the Rock-Eval oxidation stage, for a Permian carbonaceous shale sample at three different weights. See text for detailed discussion

Table 3.6 Rock-Eval oxidation results for a carbonaceous shale sample from India at different sample weights

Sample type	Weight (mgs)	TOC (wt%)	RC (wt%)	S4CO$_2$ (mg CO$_2$/g rock)	S5 (mg CO$_2$/g rock)	Oxidation mineral carbon (wt%)
Carbonaceous shale	8.51	26.67	25.41	428.61	139.49	3.8
	20.25	21.13	20.70	272.35	328.56	8.96
	45.55	14.62	14.25	209.79	546.89	14.92

The S4CO$_2$ and S5 peaks for this sample are displayed in Fig. RE16

mineral, from the Jharia basin, India, at three different sample weights of 8.51, 20.25, and 45.55 mgs sample weight. The shale sample analysis results suggest a reduction in TOC content with increasing sample weights. For the 8.51 mgs weight fraction, the TOC and S4CO$_2$ contents are 26.67 wt% and 428.61 mg CO$_2$/g rock, respectively (Fig. 3.16a). At higher weights of 20.25 and 45.55 mg, the TOC contents are 21.13 and 14.62 wt%, respectively. Correspondingly, the S4CO$_2$ measured contents are 272.35 and 209.79 mg CO$_2$/g rock for the 20.25 and 45.55 mg weight-fractions, respectively. Table 3.6 lists the different measurements recorded for this sample at the three different weight fractions.

The decline in TOC and S4CO$_2$ in the higher sample weights is caused by the elution of organic-CO$_2$ over a wider temperature range that extends beyond the defined boundary between the S4 and S5 peaks. For the 8.51 mg weight-fraction, the CO$_2$ from organic-matter is almost completely liberated below 650 °C with only a small portion of the curve extends beyond the designated 650 °C boundary between the S4 and S5 peaks. With increasing sample weights, a greater portion of the organic-CO$_2$ curve extends beyond the designated boundary between the S4 and S5 peaks, and is therefore misleadingly recorded as part of the S5 peak. This portion of the organic-CO$_2$ curve (shaded in black, in Fig. 3.17) is, consequently, undercounted for the TOC calculation, because it is attributed to the S5 peak. Correspondingly, the S5 value and oxidation mineral carbon content are misleadingly recorded to increase with increasing sample weights (Table 3.6). For analysis of bulk rocks or shales, Behar et al. (2001) and Vinci Technologies (2003) suggest using 50–70 mgs of sample sizes. The foregoing analyses clearly indicate that for organic-matter rich sediments, sample sizes should be reduced to provide more meaningful results. The S4CO$_2$ oxidation graphics of the carbonaceous shale at the lower weight-fraction (i.e., smaller sample size) (Fig. 3.17a), when compared to the S4CO$_2$ peaks recorded for the IFP and JR-1 standards (Fig. 3.16), is indicative of only organic CO$_2$ being released, which is consistent with the absence of carbonate minerals in that sample. However, for organic-rich shales containing carbonate minerals, at higher sample weights, the delineation between organic and inorganic CO$_2$ might become difficult using the S4 and S5 peak data. Thus the S4CO$_2$ and S5 oxidation peak magnitudes and

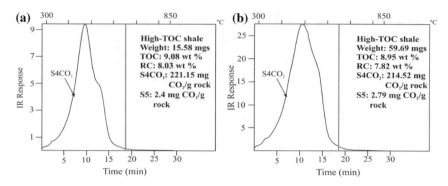

Fig. 3.18 S4CO$_2$ and S5 graphics from the Rock-Eval oxidation stage, for a Permian High-TOC shale sample from India at two different weights. See text for detailed discussion

shapes should be critically monitored for organic-rich samples, and sample weights should be kept lower to improve the reliability and consistency of the data.

For shales with lower TOC contents (<10 wt%), it is considered to be reasonable to use sample weights within the IFP specified limits of 50–70 mgs, as minimal variations are observed over a wide range of sample weights (Hazra et al. 2017). For example, for a type III kerogen-bearing Upper Permian Raniganj Formation high-TOC shale sample (~9 wt% TOC) from Raniganj basin, India, the S4CO$_2$, S5, RC, TOC, and oxidation mineral carbon displayed insignificant variations with increasing sample weight (Fig. 3.18). While at 15.58 mgs sample weight, the TOC content is measured at 9.08 wt%, at 59.69 mgs sample weight the TOC is measured at 8.95 wt%. Similar to the carbonaceous shale sample shown in Fig. 3.17, the high-TOC shale sample can also be observed to be devoid of any carbonate minerals (Fig. 3.18).

The Rock-Eval S5 peak magnitude and the calculated mineral carbon (MinC) parameter is also of significance, as it helps to establish the carbonate mineral concentration in a sample. Behar et al. (2001) observed strong positive relationships between the Rock-Eval derived MinC, and the CO$_2$-loss determined through acidimetry and decarbonation techniques. Based on this observation, Pillot et al. (2014b) identified and quantified carbonate mineral types utilizing their temperatures of decomposition, using the Rock-Eval 6 instrument. However, the temperature-ranges of dissociation of different carbonate minerals often overlap with each other making it extremely difficult to distinguish one carbonate mineral from another. Using pure carbonate-minerals and their mixtures, Pillot et al. (2014b) identified that with increasing temperatures, Cu-bearing carbonates decompose first, followed by Fe-bearing, Mg-bearing, Mn-bearing, and Ca-bearing carbonates. Identifying the presence of carbonates can be specifically significant when studying igneous intrusion affected organic-rich rocks and adjoining horizons, as a host of carbonate minerals are known to form secondarily due to re-deposition/condensation of CO$_2$ and CO produced due to interaction between magma and organic-matter (Singh et al. 2007, 2008).

Jiang et al. (2017) compared XRD mineralogical data and Rock-Eval MinC (%) of several Canadian shale samples, across different geological ages. They observed a strong positive relationship between MinC (%) and XRD-derived carbonate mineral content, and suggested that the MinC can be reliably used as a proxy to predict carbonate mineral content of shales.

References

Behar F, Vandenbroucke M (1987) Chemical modelling of kerogens. Org Geochem 11:15–24

Behar F, Kressmann S, Rudkiewicz JL, Vandenbroucke M (1992) Experimental simulation in a confined system and kinetic modelling of kerogen and oil cracking. Org Geochem 19:173–189

Behar F, Beaumont V, De B. Penteado HL (2001) Rock-Eval 6 technology: performances and developments. Oil Gas Sci Technol Rev Inst Fr Pet Energy Nouv 56:111–134

Carvajal-Ortiz H, Gentzis T (2015) Critical considerations when assessing hydrocarbon plays using Rock-Eval pyrolysis and organic petrology data: data quality revisited. Int J Coal Geol 152:113–122

Chen Y, Mastalerz M, Schimmelmann A (2012) Characterization of chemical functional groups in macerals across different coal ranks via micro-FTIR spectroscopy. Int J Coal Geol 104:22–33

Dayal AM, Mani D, Madhavi T, Kavitha S, Kalpana MS, Patil DJ, Sharma M (2014) Organic geochemistry of the Vindhyan sediments: implications for hydrocarbons. J Asian Earth Sci 91:329–338

Dembicki H Jr (2017) Practical petroleum geochemistry for exploration and production. Elsevier, 342p. ISBN: 9780128033500

Di Giovanni C, Disnar JR, Bichet V, Campy M, Guillet B (1998) Geochemical characterization of soil organic matter and variability of a postglacial detrital organic supply (Chaillexon Lake, France). Earth Surf Proc Land 23:1057–1069

Disnar JR, Guillet B, Keravis D, Di Giovanni C, Sebag D (2003) Soil organic matter (SOM) characterization by Rock-Eval pyrolysis: scope and limitations. Org Geochem 34:327–343

Espitalié J, Bordenave ML (1993) Rock-Eval pyrolysis. In: Bordenave ML (ed) Applied petroleum geochemistry. Editions Technip, Paris, pp 237–261

Espitalié J, Laporte JL, Madec M, Marquis F, Leplat P, Pauletand J, Boutefeu A (1977) Methoderapide de caracterisation des roches meres, de leur potential petrolier et de leu degred'evolution. Inst Fr Pét 32:23–42

Espitalié J, Deroo G, Marquis F (1985) La pyrolyse Rock-Eval et ses applications. Première partie. Rev Inst Fr Pét 40:73–89

Espitalié J, Deroo G, Marquis F (1986) La pyrolyse Rock-Eval et ses applications. Troisièmepartie. Inst Fr Pét 41:73–89

Ghori KAR (1998) Petroleum source-rock potential and thermal history of the Officer Basin, Western Australia: Western Australia Geological Survey, Record 1998/3, 52p

Guo YT, Bustin RM (1998) Micro-FTIR spectroscopy of liptinite macerals in coal. Int J Coal Geol 36:259–275

Hakimi MH, Ahmed AF, Abdullah WH (2016) Organic geochemical and petrographic characteristics of the Miocene Salif organic-rich shales in the Tihama Basin, Red Sea of Yemen: implications for paleoenvironmental conditions and oil-generation potential. Int J Coal Geol 154–155:193–204

Hazra B, Varma AK, Bandopadhyay AK, Mendhe VA, Singh BD, Saxena VK, Samad SK, Mishra DK (2015) Petrographic insights of organic matter conversion of Raniganj basin shales, India. Int J Coal Geol 150–151:193–209

Hazra B, Dutta S, Kumar S (2017) TOC calculation of organic matter rich sediments using Rock-Eval pyrolysis: critical consideration and insights. Int J Coal Geol 169:106–115

Hazra B, Wood DA, Kumar S, Saha S, Dutta S, Kumari P, Singh AK (2018) Fractal disposition and porosity characterization of lower Permian Raniganj Basin Shales, India. J Nat Gas Sci Eng 59:452–465

Hunt JM (1996) Petroleum geochemistry and geology, 2nd edn. W.H. Freeman and Company, New York, p 743

Inan S, Yalçin MN, Mann U (1998) Expulsion of oil from petroleum source rocks: inferences from pyrolysis of samples of unconventional grain size. Org Geochem 29(1):45–61

Jarvie DM (2012) Shale resource systems for oil and gas: part 1—shale–gas resource systems. In: Breyer JA (ed), Shale reservoirs—giant resources for the 21st Century. AAPG Memoir 97, pp 69–87

Jiang C, Chen Z, Lavoie D, Percival JB, Kabanov P (2017) Mineral carbon MinC (%) from Rock-Eval analysis as a reliable and cost-effective measurement of carbonate contents in shale source and reservoir rocks. Mar Petrol Geol 83:184–194

Jüntgen H (1984) Review of the kinetics of pyrolysis and hydropyrolysis in relation to the chemical constitution of coal. Fuel 63:731–737

Katz BJ (1983) Limitations of "Rock-Eval" pyrolysis for typing organic matter. Org Geochem 4:195–199

Kotarba M, Clayton J, Rice D, Wagner M (2002) Assessment of hydrocarbon source rock potential of Polish bituminous coals and carbonaceous shales. Chem Geol 184:11–35

Lafargue E, Espitalié J, Marquis F, Pillot D (1998) Rock-Eval 6 applications in hydrocarbon exploration, production, and soil contamination studies. Inst Fr Pét 53:421–437

Mani D, Patil DJ, Dayal AM, Prasad BN (2015) Thermal maturity, source rock potential and kinetics of hydrocarbon generation in Permian shales from the Damodar Valley basin, Eastern India. Mar Pet Geol 66:1056–1072

Pan S, Horsfield B, Zou C, Yang Z (2016) Upper Permian Junggar and Upper Triassic Ordos lacustrine source rocks in Northwest and Central China: organic geochemistry, petroleum potential and predicted organofacies. Int J Coal Geol 158:90–106

Paul S, Sharma J, Singh BD, Saraswati PK, Dutta S (2015) Early Eocene equatorial vegetation and depositional environment: biomarker and palynological evidences from a lignite-bearing sequence of Cambay Basin, western India. Int J Coal Geol 149:77–92

Peters KE (1986) Guidelines for evaluating petroleum source rock using programmed pyrolysis. AAPG Bull 70:318–386

Peters KE, Cassa MR (1994) Applied source rock geochemistry. In: Magoon LB, Dow WG (eds) The petroleum system—from source to trap, AAPG Memoir, vol 60, pp 93–120

Pillot D, Letort G, Romero-Sarmiento MF, Lamoureaux-Var V, Beaumont V, Garcia B (2014a) Procédé pour l'évaluation d'au moins une caractéristique pétrolière d'un échantillon de roche. Patent 14/55.009

Pillot D, Deville E, Prinzhofer A (2014b) Identification and quantification of carbonate species using Rock-Eval pyrolysis. Oil Gas Sci Technol Rev IFP 69(2):341–349

Romero-Sarmiento M-F, Pillot D, Letort G, Lamoureux-Var V, Beaumont V, Huc A-Y, Garcia B (2016) New Rock-Eval method for characterization of unconventional shale resource systems. Oil Gas Sci Technol 71:37

Saenger A, Cecillon L, Sebag D, Brun JJ (2013) Soil organic carbon quantity, chemistry and thermal stability in a mountainous landscape: a Rock-Eval pyrolysis survey. Org Geochem 54:101–114

Sebag D, Disnar JR, Guillet B, Di Giovanni C, Verrecchia EP, Durand A (2006) Monitoring organic matter dynamics in soil profiles by 'Rock–Eval pyrolysis': bulk characterization and quantification of degradation. Eur J Soil Sci 57:344–355

Singh AK, Singh MP, Sharma M, Srivastava SK (2007) Microstructures and microtextures of natural cokes: a case study of heat-altered coking coals from the Jharia Coalfield, India. Int J Coal Geol 71:153–175

Singh AK, Singh MP, Sharma M (2008) Genesis of natural cokes: Some Indian examples. Int J Coal Geol 75:40–48

Sykes R, Snowdon LR (2002) Guidelines for assessing the petroleum potential of coaly source rocks using Rock-Eval pyrolysis. Org Geochem 33:1441–1455

van Krevelen DW (1961) Coal: typology—chemistry—physics—constitution, 1st edn. Elsevier, Amsterdam, p 514

Varma AK, Hazra B, Samad SK, Panda S, Mendhe VA (2014) Methane sorption dynamics and hydrocarbon generation of shale samples from West Bokaro and Raniganj basins, India. J Nat Gas Sci Eng 21:1138–1147

Varma AK, Hazra B, Chinara I, Mendhe VA, Dayal AM (2015) Assessment of organic richness and hydrocarbon generation potential of Raniganj basin shales, West Bengal, India. Mar Pet Geol 59:480–490

Varma AK, Mishra DK, Samad SK, Prasad AK, Panigrahi DC, Mendhe VA, Singh BD (2018) Geochemical and organo-petrographic characterization for hydrocarbon generation from Barakar Formation in Auranga Basin, India. Int J Coal Geol 186:97–114

Vinci Technologies (2003) Rock-Eval 6 operator manual. Vinci Technologies, France

Wagner R, Wanzl W, van Heek KH (1985) Influence of transport effects on pyrolysis reaction of coal at high heating rates. Fuel 64:571–573

Wood DA, Hazra B (2018) Pyrolysis S2-peak characteristics of Raniganj shales (India) reflect complex combinations of kerogen kinetics and other processes related to different levels of thermal maturity. Adv Geo-Energy Res 2(4):343–368

Zhang S, Liu C, Liang H, Wang J, Bai J, Yang M, Liu G, Huang H, Guan Y (2018) Paleoenvironmental conditions, organic matter accumulation, and unconventional hydrocarbon potential for the Permian Lucaogou Formation organic-rich rocks in Santanghu Basin, NW China. Int J Coal Geol 185:44–60

Chapter 4
Matrix Retention of Hydrocarbons

During any open-system anhydrous programmed pyrolysis experiments such as the Rock-Eval technique, the mineral matrix components present within the rock can significantly affect the pyrolysis results. Several early studies (Horsfield and Douglas 1980; Espitalié et al. 1980) identified the key impacts of mineral matrix on the S2 peak generated by Rock-Eval analysis. Essentially, the retention of petroleum fluids within the mineral matrix leads to shape alteration of the S2 peak. The exact nature of these impacts on the S2 peak depends on host-rock matrix mineralogy and texture.

Katz (1983) isolated type I kerogen from Green River Shale (USA) and mixed it with calcite and calcium-rich montmorillonite (clay). Greater quantities of petroleum were released/generated from the carbonate matrix in comparison to the argillaceous matrix in the pyrolysis tests conducted on that material. Moreover, there was an increase in hydrogen index (HI) with increasing TOC content for a suite of samples containing similar kerogen components. Those results suggest that as kerogen content in a rock increases, it's potential to liberate and expel petroleum fluids also increases, overwhelming the petroleum-fluid-retention capacity of the rock matrix.

Argillaceous matrices within shales can retain greater amounts of the petroleum fluids generated during pyrolysis delaying or inhibiting their release and onward migration. Using variable oil shale-minerals and kerogen-mineral mixtures, Espitalié et al. (1980) documented that petroleum fluids retained during pyrolysis are mostly caused by clay minerals, with illite showing the maximum effects. Such effects can also cause distortions (typically an increase) in Rock-Eval T_{max} values leading to improper, and/or potentially misleading, source-rock assessments (Espitalié et al. 1984; Peters 1986). In most instances studied clay minerals present within shales are identified as the specific sites where the petroleum fluids, released from the maturing kerogens, are retained. However, different clay minerals have different microporosities and pore size distribution (Ross and Bustin 2009; Ji et al. 2012) making their petroleum-fluid retention capacities variable.

As the organic-facies present within a formation can also significantly affect the S2 versus TOC relationship, it is important to take these into account when considering the matrix-retention effects (Hazra et al. 2018a). Type I and type II kerogen generates much greater volumes of hydrocarbons during pyrolysis than type III and

© Springer Nature Switzerland AG 2019   51
B. Hazra et al., *Evaluation of Shale Source Rocks and Reservoirs*,
Petroleum Engineering, https://doi.org/10.1007/978-3-030-13042-8_4

type IV kerogen, affecting Rock-Eval estimates (see Sect. 2.2 for further discussions). Type I and II kerogens tend to saturate the rock matrix with petroleum fluids even at lower percentage concentration, and the variable impacts of matrix-retention tend to have minimal impacts on the pyrolysis curves (Katz 1983). On the other hand, type III kerogens, due to their lower hydrogen contents, generate lower volumes of pyrolyzates which do not saturate the rock matrix even at higher concentrations, and thereby the effect of matrix-retention is more pronounced in the pyrolysis curves (Katz 1983). The mineral matter in addition to retention of petroleum fluids can also affect the source-rock kinetic parameters (Dembicki 1992). Using kerogen extracted from a Kimmeridgian (Jurassic) black shale sample from Dorset, England, and mixing it with different minerals in varying proportions, Dembicki (1992) observed distinct behaviors and effects of clay minerals compared to non-clay minerals on kinetic parameters. At low-TOC content, the non-clay minerals (viz. calcite, dolomite, quartz) were observed to retain some petroleum fluids released under the S2 peak causing an increase in the activation energy distribution required to fit the recorded S2 peak shape, compared to the pyrolysis result of 100% isolated kerogen. When the TOC content of the kerogen and non-clay minerals mixture was increased, the activation energy was observed to approach that of 100% isolated kerogen. In contrast, when montmorillonite and kaolinite clays were mixed with kerogen, at low-TOC levels, the activation energy was observed to be lower compared to that of 100% isolated kerogen, pointing towards catalytic effects of these clays on petroleum fluid generation. With increasing TOC content of the mixtures, the activation energy distributions were observed to be minimized for mixtures containing montmorillonite, indicating deactivation of catalytic sites. On the other hand, for the mixtures containing kaolinite, even with increasing TOC-contents, constant activation energy distributions were observed, indicating catalytic sites remain active during the petroleum generation process.

Ideally, when processing Rock-Eval data for analysis, a linear regression curve is generated for the S2 versus TOC relationship, facilitating the calculation of HI (Langford and Blanc-Valleron 1990). The best-fit line of that regression should pass through the origin, unless matrix causes retention of petroleum fluids impacting the pyrolysis results. Under such circumstances, the best-fit line for S2 versus TOC intercepts the TOC-axis, instead of passing through the origin (Peters 1986; Espitalié et al. 1980; Langford and Blanc-Valleron 1990).

Figure 4.1 plots S2 versus TOC data of 101 shale samples from two Indian Gondwana basins viz. Raniganj and Auranga studied by Hazra et al. (2018b), Mendhe et al. (2018a, b), Varma et al. (2018). These shales are predominantly composed of type III-IV kerogens, mixed with type II kerogens in some samples. The best-fit line for these shales, cuts the TOC-axis at 49 wt%, indicating that, for this Permian suite of rocks at least, 49 wt% TOC is required before expulsion of petroleum fluids can occur. Langford and Blanc-Valleron (1990) suggested that in such cases, the associated negative intercept on the S2 axis can be more informative and useful. That negative intercept is less dependent on the character of the organic matter compared to TOC. Consequently, that negative S2 intercept can be used to apply an approximate correction for mineral matrix effects. In Fig. 4.1, the best-fit line intersects the

S2-axis on the negative side at 4.231 which indicates that the adsorptive capacity of 1 gm of rock for this suite is 4.231 mg HC released under S2 curve of Rock-Eval.

The S2 versus TOC relationship for type I kerogen-bearing Permian shales from Santanghu Basin, China reveal a different matrix effect (Zhang et al. 2018). Those samples reveal a much smaller intercept on the TOC-axis associated with a more significant negative intercept on the S2 axis (Fig. 4.2). The best-fit line for the Santanghu-basin samples intercepts the TOC-axis at 0.859 wt% (i.e., lower than the Permian shales from India), and intercepts the S2-axis on the negative side at −8.21 (i.e., a more significant magnitude than displayed by the Permian shales from India). These results indicate that although matrix retention is significant for the Permian shales from Santanghu-basin shale samples, due to their higher HI values (421–918 mg HC/g TOC) and presence of high pyrolyzate-yielding type I kerogen, the matrix-retention effects are easily overwhelmed, resulting in a lower positive intercept on the TOC-axis. In contrast, for the dominantly type III–IV kerogen bearing Permian shales from India, due to their inherent lower petroleum-fluid yields, the positive intercept on the TOC-axis is higher.

Similar results were also observed by Hazra et al. (2018a) when comparing the S2 versus TOC relationships for type III-IV kerogen-bearing Permian shales from India and type I-II kerogen-bearing shales from the Paleocene-Eocene Çamalan Formation, Nallıhan–Turkey studied by Sari et al. (2015). While the negative S2 intercept was observed to be similar for those two shale-suites, the positive intercept on the TOC-axis was observed to be much smaller for the Paleocene-Eocene type I-II kerogen-bearing shales than the Permian type III-IV shales. This difference is similar to the observations displayed for the two sample suites compared in Figs. 4.1 and 4.2. It

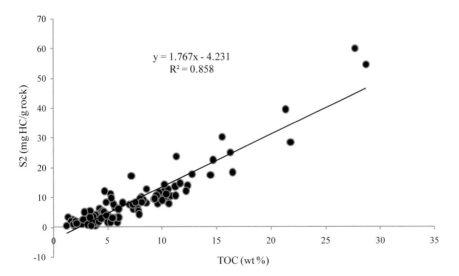

Fig. 4.1 Cross-plot showing S2-TOC relationship for Permian shales from India [data compiled from Hazra et al. (2018b); Mendhe et al. (2018a, b); Varma et al. (2018)]

leads to the interpretation of lower or negligible concentrations of inert organic-matter within the high pyrolyzate-yielding shales from both Turkey and China. Similar smaller TOC-axis intercepts are also observed for type I-II kerogen-bearing shales from different parts of the world (Fig. 4.3).

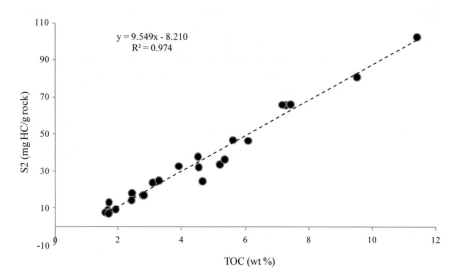

Fig. 4.2   Cross-plot showing S2-TOC relationship for Permian shales from Santanghu Basin, China (Zhang et al. 2018)

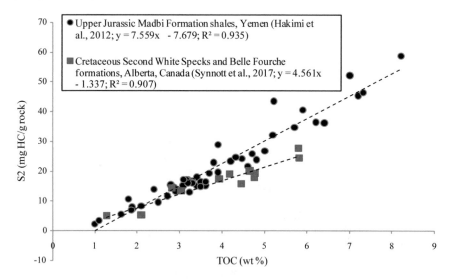

Fig. 4.3   Cross-plot showing S2-TOC relationship for Upper Jurassic Madbi Formation shales, Yemen (Hakimi et al. 2012), and Cretaceous Second White Specks and Belle Fourche formations, Alberta, Canada (Synnott et al. 2017)

The presence of inert organic matter or type IV kerogen within shales can significantly impact the S2 versus TOC relationship for a suite of rocks (Cornford 1994; Cornford et al. 1998; Dahl et al. 2004). As type IV kerogen has very limited, if any, petroleum-generation capacity, its presence in relatively greater concentration in some samples from a suite a rocks, can impact their intercept along the TOC-axis. Consequently, when analysts opt for source-rock matrix-correction using graphical analysis of the S2 versus TOC relationships for samples containing admixed kerogen types, the effects of the inert organic-matter should be corrected first. It is desirable for the correction of the TOC curve to only include reactive TOC. To achieve this, it is necessary to initially deduct the inert TOC component. Removing the inert TOC should lead to a reduction in the TOC intercept value and a less negative S2 intercept value making the matrix retention correction more meaningful (Hazra et al. 2018a).

Using optical microscopic techniques, analysts can easily identify the volume percentages of inert and reactive organic-matter, and then use that information to ensure that only the impact of reactive TOC content on S2 versus TOC relationship (Hazra et al. 2018a) is considered for matrix retention corrections. Hazra et al. (2018a), based on previous studies, considered 70 vol. % of the inertinites present in their sample suite to be non-reactive. Applying that factor and the ratio of organic matter to TOC they calculated the reactive-TOC component for a suite of Permian shales from India. While the initially non-corrected Permian shales (i.e., including the inert-TOC) showed higher TOC-axis and S2-axis intercepts, the inert-corrected shales showed smaller TOC- and S2-axis intercepts. The latter, corrected intercepts, are interpreted to be more representative of the actual amounts of petroleum-fluids retained by the shale-matrix. Using this intercept on the S2-axis for that suite of shales, and the reactive-TOC component, they determined a retention-corrected hydrogen index (HI) for that sample suite. As a rule, they recommended that in order to establish meaningful retention-corrected HI values, only samples with a specific range of TOC contents and with consistent kerogen types should be used to define the underlying S2 versus TOC relationship used for matrix-retention correction calculations. This may mean generating more than one S2 versus TOC curves to distinguish sub-groups of samples with distinctive kerogen distributions.

References

Cornford C (1994) The Mandal-Ekofisk(!) Petroleum system in the Central Graben of the North Sea. In: Magoon LB, Dow WG (eds) From source to trap. AAPG Memoir 60, Tulsa, pp 537–571

Cornford C, Gardner P, Burgess C (1998) Geochemical truths in large data sets I: geochemical screening data. Org Geochem 29:519–530

Dahl B, Bojesen-Koefoed J, Holm A, Justwan H, Rasmussen E, Thomsen E (2004) A new approach to interpreting Rock-Eval S2 and TOC data for kerogen quality assessment. Org Geochem 35:1461–1477

Dembicki H Jr (1992) The effects of the mineral matrix on the determination of kinetic parameters using modified Rock Eval pyrolysis. Org Geochem 18:531–539

Espitalié J, Madec M, Tissot B (1980) Role of mineral matrix in kerogen pyrolysis: influence on petroleum generation and migration. AAPG Bull 4(1):59–66

Espitalié J, Makadi KS, Trichet J (1984) Role of the mineral matrix during kerogen pyrolysis. Org Geochem 6:365–382

Hakimi MH, Abdullah WH, Shalaby MR (2012) Geochemical and petrographic characterization of organic matter in the Upper Jurassic Madbi shale succession (Masila Basin, Yemen): Origin, type and preservation. Org Geochem 49:18–29

Hazra B, Wood DA, Varma AK, Sarkar BC, Tiwari B, Singh AK (2018a) Insights into the effects of matrix retention and inert carbon on the petroleum generation potential of Indian Gondwana shales. Mar Pet Geol 91:125–138

Hazra B, Wood DA, Kumar S, Saha S, Dutta S, Kumari P, Singh AK (2018b) Fractal disposition and porosity characterization of lower Permian Raniganj basin shales, India. J Nat Gas Sci Eng 59:452–465

Horsfield B, Douglas AG (1980) The influence of minerals on the pyrolysis of kerogens: Geochimica et Choschimica Acta 44:1119–1131

Ji L, Zhang T, Milliken KL, Qu J, Zhang X (2012) Experimental investigation of main controls to methane adsorption in clay-rich rocks. Appl Geochem 27:2533–2545

Katz BJ (1983) Limitations of "Rock-Eval" pyrolysis for typing organic matter. Org Geochem 4:195–199

Langford FF, Blanc-Valleron MM (1990) Interpreting rock-Eval pyrolysis data using graphs of pyrolyzable hydrocarbons versus total organic carbon. AAPG Bull 74:799–804

Mendhe VA, Mishra S, Varma AK, Kamble AD, Bannerjee M, Singh BD, Sutay TM, Singh BD (2018a) Geochemical and petrophysical characteristics of Permian shale gas reservoirs of Raniganj Basin, West Bengal India. Int J Coal Geol 188:1–24

Mendhe VA, Kumar S, Kamble AD, Mishra S, Varma AK, Bannerjee M, Mishra VK, Sharma S, Buragohain J, Tiwari B (2018b) Organo-mineralogical insights of shale gas reservoir of Ib-River Mand-Raigarh Basin India. J Nat Gas Sci Eng 59:136–155

Peters KE (1986) Guidelines for evaluating petroleum source rock using programmed pyrolysis. AAPG Bull 70:318–386

Ross DJK, Bustin RM (2009) The importance of shale composition and pore structure upon gas storage potential of shale gas reservoirs. Mar Pet Geol 26:916–927

Sari A, Moradi AV, Akkaya P (2015) Evaluation of source rock potential, matrix effect and applicability of gas oil ratio potential factor in Paleocene - Eocene bituminous shales of Çamalan Formation, Nallıhan—Turkey. Mar Pet Geol 67:180–186

Synnott DP, Dewing K, Sanei H, Pedersen PK, Ardakani OH (2017) Influence of refractory organic matter on source rock hydrocarbon potential: a case study from the Second White Specks and Belle Fourche formations, Alberta Canada. Mar Petrol Geol 85:220–232

Varma AK, Mishra DK, Samad SK, Prasad AK, Panigrahi DC, Mendhe VA, Singh BD (2018) Geochemical and organo-petrographic characterization for hydrocarbon generation from Barakar Formation in Auranga Basin India. Int J Coal Geol 186:97–114

Zhang S, Liu C, Liang H, Wang J, Bai J, Yang M, Liu G, Huang H, Guan Y (2018) Paleoenvironmental conditions, organic matter accumulation, and unconventional hydrocarbon potential for the Permian Lucaogou Formation organic-rich rocks in Santanghu Basin NW China. Int J Coal Geol 185:44–60

Chapter 5
Kerogen's Potential to Be Converted into Petroleum: Reaction Kinetics and Modelling Thermal Maturity Plus Petroleum Transformation Processes

5.1 Kerogen and the Significance of Its Biogenic and Thermogenic Evolution

It is the burial of organic-rich fine-grained sediments deposited at the earth's surface that leads initially to the formation of the organic macerals in coals and shales made up mainly of microscopic plants and animals. As burial depth increases these macerals are progressively transformed by heat and pressure into the organic mineral kerogen that is not easily dissolved in organic acids. The composition of kerogen, in particular its hydrogen/carbon ratio is determined by the type and source of the organic material and the environment in which it was originally deposited and preserved. The four types of kerogen commonly distinguished (type 1—oil prone—lacustrine/land-based origin; type II—oil and gas prone—marine origin; type III—gas prone—terrestrial origin; type IV—barren—varied origin but mainly terrestrial) determine the nature of the petroleum that will be generated as the sediments containing them are buried evermore deeply (Fig. 5.1). Kerogen types I and II tend to be more hydrogen rich than kerogen types III and IV that are dominated by carbon.

Biogenic gases are produced, in large quantities in anaerobic conditions, at shallow burial depths. Microbes generating biogenic gas are most active at burial depths of <550 m (Shurr and Ridgley 2002). Natural gases produced by microbial degradation of organic material are readily distinguished from thermogenic gases because they are very dry with low ethane and other natural gas liquid contents and with distinguishably light carbon isotopes in the methane (Whiticar 1994). Biogenic gas has two distinct origins: (1) primary—produced by degradation of organic material in very shallow sediments (less than about 500 m); and (2) secondary—produced by the biodegradation of shallow reservoirs of thermogenic oil and wet gas. Both sources can produce commercially viable quantities of reservoir gas, which, subsequent to its generation, can be buried to depths below the base of the biogenic gas generation window and in some cases become mixed with thermogenic gas. More than twenty percent of all reservoired gas around the world is believed to be of biogenic origin (Rice and Claypool 1981). Most biogenic methane is produced in an

© Springer Nature Switzerland AG 2019
B. Hazra et al., *Evaluation of Shale Source Rocks and Reservoirs*,
Petroleum Engineering, https://doi.org/10.1007/978-3-030-13042-8_5

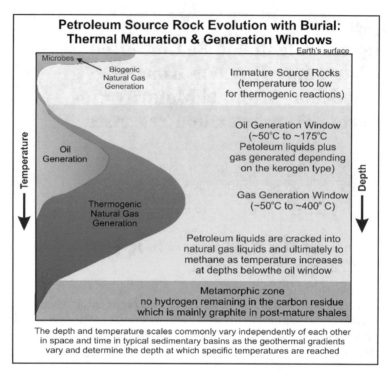

Fig. 5.1 Conceptual diagram describing oil and gas generation in relation to oil and gas generation windows linked to subsurface temperature intervals

anaerobic sulphur-free environment at temperatures <75 °C in which carbon dioxide is reduced by hydrogen produced by microbial actions that metabolize (ferment) the labile portion of organic matter present in a complex series of reactions (Cokar et al. 2013). Large biogenic gas reservoirs in the marine sediments of the Eastern Mediterranean (Schneider et al. 2016), in the lacustrine sediments of the Sanhu area in the Qaidam basin of NW China (Yang et al. 2012), in organic-rich shales uplifted into the biogenic zone in Alberta Canada (Cokar et al. 2013), and the biogenic origin of some gas hydrate deposits, testify to the growing commercial significance of global biogenic gas resources. The effectiveness of biogenic gas generation depends to an extent on the relationship between burial rate and geothermal gradient that determines how much time organic-rich sediments spend in the biogenic-gas-generation zone during their burial history. Its formation is difficult to model and quantify using first-order reaction kinetics.

Organic-rich sediments need to be buried more deeply beneath younger accumulating sediments in sedimentary basins to reach the temperature-related (thermogenic) oil and gas generation windows (also referred to as petroleum kitchen areas) where they become thermally mature and parts of the kerogen those sediments contain generate petroleum. Sufficient burial to reach specific temperature thresholds, time spent in different zones of the oil and gas generation windows and pressure all play roles in the quantities and rates of petroleum generated from specific petroleum sources rocks. As heat flow and geothermal gradients vary from basin to basin around the world, depending upon crustal-scale tectonic environments, and, indeed, within the same basin, it is not possible to link the oil and gas generation windows with specific depth ranges without considering the historical evolution of geothermal gradients over geological time scales within a specific basin. Burial of sediments is also not a smooth and continuous process. Many formations are subjected over their geological history to intermittent periods of rapid burial beneath an accumulating sedimentary overburden, with periods of uplift due to the erosion of overlying sediments, and periods of isothermal conditions as burial is temporarily halted. The time a sediment spends at each temperature during its burial history has an influence on how much of its kerogen is transformed into petroleum and when that transformation occurs. A formation's burial and thermal history is more significant in that regard than its current depth of burial.

As shales reach various levels of thermal maturity their kerogen(s) generates oil and/or gas (depending upon its composition). Some of the petroleum generated is expelled from the kerogen (referred to as primary migration) and resides in the limited pore-space and fractures in the tight rock formation. The remainder of the petroleum generated remains in the pore space within the kerogen, that pore space gradually increasing as thermal maturity advances and more petroleum is generated within the kerogen. Some of the petroleum residing in the matrix and fractures of the shale, typically as pressure within the formation builds, due partly to the increase in fluid present per unit volume, will be expelled from the shale formation. That expelled petroleum will make its way up-dip (it is less dense than water), or up fractures and faults, to be trapped temporarily or semi-permanently in shallower porous (or tight) rock formations (secondary migration). The fraction of petroleum expelled from a shale varies but is typically only a relatively small fraction of that generated; meaning that much of the petroleum generated remains within the shale as an extensive petroleum resource to be exploited in addition to the conventional porous reservoirs fed by the fluids expelled and transported by secondary migration routes.

The initial generation of petroleum in the thermally mature kerogens is a process termed catagenesis. However, other thermal transformation processes also impact the composition of the petroleum generated by catagenesis. In particular the larger hydrocarbon molecules making up liquid petroleum (also referred to generically as bitumen) and natural gas liquids are cracked under the pressure and temperature exerted on the formation over time to progressively transform them into smaller molecules. This cracking transformation progresses ultimately towards a methane-rich (CH_4) petroleum fluid (gas) dominating the petroleum contained within sediments that reach the deeper end of the thermogenic gas window. Shales that are very tight (i.e. very low permeability), and/or securely isolated by impermeable clay-rich top and bottom seals, may not be able to easily expel the petroleum generated from within them. As such rocks are buried more deeply, they retain most of their petroleum, but it becomes progressively more methane-rich.

The depth and thickness of shales tend to vary across a typical sedimentary basin, because subsidence and deposition rates vary spatially. Also, the depths to the boundaries of the oil and gas generation windows also tend to vary spatially due to heterogeneity in heat flow on a basin-wide scale. These variations cause the petroleum-generation potential and composition of the petroleum present within a shale to vary significantly with location and depth across a typical basin. This means that in order to find the petroleum "sweet spots" within an organic-rich shale distributed across vast tracts of a sedimentary basin it is necessary to map in detail the organic content, kerogen type, shale thickness and depth in relation to the oil and gas generation window temperature intervals. Armed with such information, together with burial history information it is possible to quantify the level of thermal maturity of specific shale layers and establish how much petroleum they are likely to have generated.

How much petroleum is generated per unit of organic carbon from a shale will depend upon its kerogen type (e.g. hydrogen index) and the level of thermal maturity it has reached. Figure 5.2 highlights that the majority of the organic carbon present in organic-rich shales is non-productive (i.e., it is non-generative organic carbon, NGOC) yielding no petroleum. Typically, less than 40% of the organic carbon present (linked to its hydrogen index value) in a thermally-immature shales has potential to produce petroleum (i.e., it is generative organic carbon, GOC) yielding petroleum. In a post-mature region of the same shale formation (i.e., one within the deeper parts of the thermogenic gas window or deeper) less than 2% of the residual carbon present is likely to be GOG. This transformation of the TOC reflects the changes that are occurring to the kerogen as petroleum generation progresses. Typically, only some of the petroleum generated will remain in the shale after it has resided in thermally mature zones for significant periods. The petroleum remaining in the shale is of interest because of its shale resource (unconventional petroleum) potential; the portion of the petroleum generated, expelled via secondary migration and trapped elsewhere in other porous and tight formations is of interest because of its potential to feed conventional porous and permeable petroleum reservoirs.

5.2   Thermal Maturity Modelling of Organic-Rich Sediments

Thermal maturation modelling is required to determine the extent and timing of petroleum generation from shales and other organic-rich formations at various locations and depths in basins to establish their potential as viable conventional petroleum source rocks and/or for exploitation as unconventional oil and gas reservoirs. Reliable basin-wide analysis can be provided by thermal maturity models, if they can be meaningfully calibrated with thermal maturity measurements taken from borehole samples (e.g. Ro and T_{max} data) applying plausible burial histories and geothermal gradients to the kerogen(s) present. Such analysis can identify the sweet spots, defined

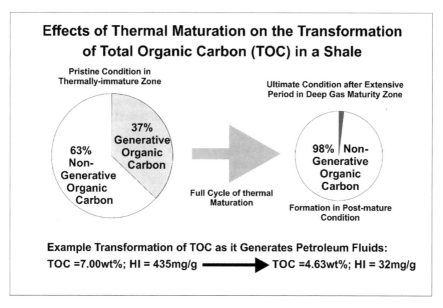

Fig. 5.2 Only a component of the TOC in a shale is capable of generating petroleum and will do so progressively as it passes though the various zones of the oil and gas generation windows. Modified after Jarvie (2014), Wood and Hazra (2017)

in terms of specific spatial coordinates and depth intervals, in which for commercial quantities of petroleum liquids wet gas and dry gas are most likely to occur.

The thermogenic transformation of kerogen to petroleum involves, in detail, a suite of multiple chemical reactions. The general approach is to simplify this when modelling to consider just the first-order reactions applying the Arrhenius equation, on the basis that such reactions dominate the primary breakdown of kerogen to petroleum fluids. Subsequent cracking and reforming processes clearly occur in the sub-surface modifying the composition of the petroleum fluids, but these are secondary to the first-order reactions.

5.3   History of Thermal Maturity Modelling of Source Rocks

Quantitative attempts to model thermal maturity of petroleum source rocks dates back to the 1970s (Lopatin 1971) with attempts to link a modelled time-temperature index (*TTI*) based on that method to vitrinite reflectance (Waples 1980). However, it soon became clear (Tissot and Espitalié 1975; Tissot and Welte 1978) that first-order reactions with rates determined by the Arrhenius equation provided more realistic scientific foundations for the processes involved in transforming of kerogen to petroleum.

Wood (1988) developed the Arrhenius-equation-based, cumulative time-temperature index ($\sum TTI_{ARR}$) method for thermal maturity modelling of petroleum source rocks. This applied the cumulative integration of the Arrhenius equation based upon a single activation energy (E = 218 kJ/mol) and pre-exponential factor ($Log_e A$ = 61.56/million years) to model thermal maturity and relate it to observed vitrinite reflectance (Ro) trends for a diverse range of burial histories. The kerogen kinetic values used in that model were taken as representative of type II and type III kerogens based on data available at that time (Fig. 5.3). Larter (1989) suggested the use of a normal distribution of activation energies to better represent the series of parallel first-order Arrhenius reactions deemed to be involved in the thermal maturation of vitrinite. Sweeney and Burnham (1990) developed that approach into their "Easy-Ro" thermal maturation model that involved applying a flexible distribution (i.e., not conforming to any specific mathematical distribution type) of twenty activation energies (E) but with a constant pre-exponential factor (A). The central value of their distribution was close to that proposed by Wood (1988). Although some questioned the kinetic validity of the multiple E values associated with a single A value (Nielsen and Barth 1991), the multiple E-constant A approach has been widely used since the 1990s (Pepper and Corvi 1995; Dieckmann 2005; Cornford 2009; Stainforth 2009). However, Wood (2017, 2018a) provide further justification and advantages for using the $\sum TTI_{ARR}$ method developed in 1988, to model single and mixed kerogens and question the validity of using distributions of activations energies assuming a constant A value for such modelling. Wood (2017) provides a more detailed summary of the history of thermal maturity modelling of source rocks and it is the $\sum TTI_{ARR}$ method that is described in more detail here.

5.4 Calculating the Cumulative Arrhenius Time Temperature Index ($\sum TTI_{ARR}$)

A convenient way to express the Arrhenius equation (Arrhenius 1889) is Eq. (5.1):

$$E = RT \log_e(A) - RT \log_e(k) \tag{5.1}$$

where

E is the activation (in kJ/mol units; although, U.S. laboratories still use kcal/mol units).
R is the universal gas constant (0.008314 kJ/mol).
T is the absolute temperature (in degrees Kelvin °K).
e the mathematical exponent.
A is the pre-exponential (or frequency) factor, expressed for burial-history modelling in geological time scale of per millions of years or for laboratory-scale models as per minute or per seconds.
k is the reaction rate of a first order reaction.

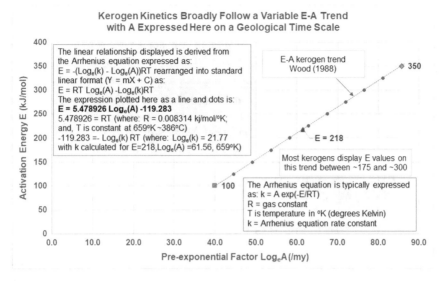

Fig. 5.3 A long-established trend (Wood 1988) of activation energy (E) versus pre-exponential (frequency) factor ($\log_e A$) for hydrocarbon reaction kinetics applying the Arrhenius equation on a geological time scale. The cumulative time-temperature thermal maturation index $\sum TTI_{ARR}$ is developed using kinetics (E = 218 kJ/mol; $\log_e A$ = 61.56) at a near central point on this trend

Wood (1988) used Eq. (5.1) to describe a linear trend of E (kJ/mol) versus $\log_e A$ showing that a number of different kerogens, based on published data at that time (e.g., Tissot and Espitalié 1975; Espitalié et al. 1977; Lewan 1985), followed that line (Fig. 5.3). Indeed, this trend was used to justify the selection of representative kerogen kinetics of (E = 218 kJ/mol) and pre-exponential factor (A = 5.4349E + 26/million years) for calculating the ($\sum TTI_{ARR}$) index.

The $\sum TTI_{ARR}$ is based on a temperature integral of the Arrhenius equation with a simple calculated time adjustment factor applied to each time interval modelled linked to the prevailing heating rate. For each time interval *TTI* is expressed in a form that can be readily calculated for geological and laboratory time scales as Eq. (5.2).

$$TTI_{ARR}\left(t_n \text{ to } t_{n+1}\left[T_n \neq T_{n+1}\right]\right) = \frac{A}{qn}\left[\frac{RT_{n+1}^2}{E + 2RT_{n+1}}e^{-\frac{E}{RT_{n+1}}} - \frac{RT_n^2}{E + 2RT_n}e^{-\frac{E}{RT_n}}\right]$$

$$(5.2)$$

where

qn is the heating rate (°C/ millions of years) for the time interval n to $n + 1$. Based on the temperatures at time points t_n and t_{n+1} an appropriate value of q_n can be calculated for each time interval modelled.

$\frac{A}{qn}$ is the time adjustment factor for the time interval t_n and t_{n+1}.

T_n is the modelled formation's temperature at time t_n; T_{n+1} is the modelled formation's temperature at time t_{n+1}.

$T_n \neq T_{n+1}$ means that the Eq. (5.2) calculation only applies when temperatures are changing across the time interval t_n and t_{n+1}.

TTI expressed in this way is a temperature integral with a specific time adjustment factor applied to each time interval modelled. As the time between t_n and t_{n+1} increases for a specific temperature difference between T_n and T_{n+1} the value of qn decreases causing the time adjustment factor A/qn to increase. In the burial conditions typically involved over geological time scales heat flow and geothermal gradients vary over time, which are easily accommodated in this TTI formulation (Eq. 5.2). For the special case where formation temperature remains constant across time interval t_n and t_{n+1} the *TTI* calculation is then simplified to Eq. (5.3):

$$TTI_{ARR}\left(t_n \text{ to } t_{n+1}\left[T_n = T_{n+1}\right]\right) = (t_n \text{ to } t_{n+1})Ae^{-E/RT_n} \tag{5.3}$$

The *Cumulative Arrhenius Time Temperature Index* $\sum TTI_{ARR}$ is then calculated as the sum of the TTI values for each interval modelled which is expressed as Eq. (5.4)

$$\sum TTI_{ARR} = \sum_{n=1}^{n=m} \text{Eq. (5.2)} \textit{ for time intervals where } T_n$$

$$\neq T_{n+1} + \sum_{n=1}^{n=m} \text{Eq. (5.3)} \textit{ for time intervals where } T_n = T_{n+1} \tag{5.4}$$

A more detailed derivation of the $\sum TTI_{ARR}$ formulations is provided in (Wood 1988, 2017). Significantly the $\sum TTI_{ARR}$, as defined, is effectively correlated to equivalent vitrinite reflectance (Ro) for the Ro value range 0.2–4.7% (Wood 2018b) using two polynomial equations for different Ro intervals (Eqs. 5.5 and 5.6).

For the vitrinite reflectance range 0.2% $\leq$ Ro < 1.1% Eq. (5.5) applies:

$$Ro_{calc}(\%) = 3E - 05x^4 + 0.0013x^3 + 0.0198x^2 + 0.1726x + 0.9612 \tag{5.5}$$

where

$x = \log_{10} \sum TTI_{ARR}$
if Ro_{calc} is calculated to be less than 0.2% by Eq. (5.5) then Ro_{calc} is fixed at 0.2.

For the vitrinite reflectance range 1.1% $\leq$ Ro $\leq$ 4.7% Eq. (5.6) applies:

$$Ro_{calc}(\%) = -0.0019x^4 + 0.023x^3 - 0.0483x^2 + 0.3318x + 0.8975 \tag{5.6}$$

if Ro_{calc} is calculated to be greater than 4.7% by Eq. (5.6) then Ro_{calc} is fixed at 4.7.
if $\log_{10} \sum TTI_{ARR}$ is calculated to be greater than 8.3 by Eq. (5.4) then Ro_{calc} is also fixed at 4.7.

These relationships between $\log_{10} \sum TTI_{ARR}$ and Ro have been tested across the full range using a number of varied burial histories (Wood 2017) and have been shown to match the calculated Ro scale proposed by the "Easy-Ro" method of Sweeney and Burnham (1990).

There are other integrals solutions derived for the Arrhenius equation that are sometimes used for thermal maturity modelling particularly those that apply constant heating rates in laboratory pyrolysis experiments and inverting kinetics from Rock-Eval pyrograms, such as Eq. (5.7) (Chen et al. 2017).

$$dx/dT \approx \frac{ART^2}{q}\left[1 - \frac{2RT}{E}\right]e^{-E/RT_{n+1}} \tag{5.7}$$

However, these are not so easily evaluated as Eq. (5.4) for multiple heating rates and complex burial histories over geological time scales.

5.5 Burial History Modelling

In order to realistically model the thermal maturity of sedimentary formations deposited many millions of years ago it is firstly necessary to reconstruct their burial and thermal histories. This is a complex task in its own right, typically with a number of uncertainties involved, particularly in cases where the burial history involves periods of uplift and erosion and the thickness of the eroded sections needs to be estimated. Estimating paleo-heat flow and geothermal gradient variations over space and time within a basin also commonly involves significant uncertainty.

In practice, the only information that is available comes from a few boreholes with sporadic geochemistry and measured thermal maturity data available, e.g., vitrinite reflectance (and other geochemical maturity measures) and Rock-Eval analysis for the organic-rich formations which typically only exist sporadically through the geological sections drilled. This is also often supplemented by outcrop data, regional-scale depth to basement mapping, today's heat flow profiles and mineralogical indicators of the thermal history available for some formations (e.g., fission tracks in apatite, Huntsberger and Lerche 1987; Donelick et al. 2005) across the basin of interest.

Multi-dimensional burial and thermal history modelling has been conducted for several decades (Nunn et al. 1984) highlighting the wide-range of burial histories that different types of petroleum-producing basins have been subjected to (Wood 1988). Graphical displays illustrating the evolution of the thermal maturity of geological sequences through time are now routinely used as part of wider basin analysis studies (e.g., He and Middleton 2002; Mohamed et al. 2016; Yang et al. 2017). Figure 5.4 shows an example burial and thermal maturity history graphic constructed for a subsidence history analogous to that of the central/deepest part of the 150-million-year old Melut rift basin in Sudan (Mohamed et al. 2016; Wood 2018a). This basin

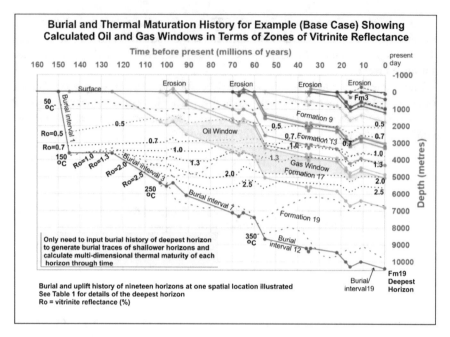

Fig. 5.4 Multi-dimensional burial and thermal maturation history modelled for a complex burial history of a 150-million-year old rift basin with four periods of erosion (Wood 2018a). This information is of particular value when identifying the timing of oil and gas generation from specific source-rock intervals for conventional oil and gas play exploration. It is also relevant to unconventional shale oil and gas reservoir characterization

involves a complex burial and thermal history with several periods of uplift and erosion interspersed with periods of rapid burial and variable geothermal gradients.

The burial history displayed in Fig. 5.4 is based on the data listed in Table 5.1. The row in Table 5.1 referenced as "End Interval 1" represents the depth to the base of the first sedimentary formation deposited at the end of the first period of deposition (i.e. 145 million years ago). The final row in Table 5.1 ("end Interval 19") represents the depth to the base of the first sedimentary formation deposited at the present day after 19 periods of burial and/or uplift and/or no deposition. With the burial history information provided for just the deepest formation (Table 5.1) it is possible to reconstruct the burial history for the younger formations deposited above it (as shown in Fig. 5.4).

To convert a multi-dimensional burial into a thermal maturation reconstruction over geological time, such as the example provided (Fig. 5.4, Table 5.1) to evaluate the petroleum generation potential of a specific point within a basin, three analytical steps are required:

Step 1: Apply a robust burial history algorithm capable of incorporating periods of uplift and erosion and variable geothermal gradients. Typically, this should be capable

Table 5.1 Burial history and thermal maturation model providing calculated Ro and $\sum TTI_{ARR}$ values for the deepest horizon in the complex burial history of a 150-million-year old rift basin with four periods of erosion illustrated in Fig. 5.4 (modified from Wood 2018a)

Burial and Thermal History Model for an Example Deep Rift Basin Sedimentary Sequence Since its Deposition											
	Multi-dimensional Burial and Thermal History Generated from Inputing the Deepest Formation as a Burial History						Arrhenius Equation Time-Temperature Index (TTI) Single Activation Energy Model E= 218 kj/mol (52.1 kcal/mol) A=5.45 E26 /m.y.				
Burial Interval	Depth to Formation Tops (metres)	Time / Age (millions of years - m.y.)	Burial Rate dz/dt (m/m.y.)	Temperature Gradient (°C /m)	Heating Rate dT/dt (°C/m.y.)	Temperature (°C)	Thermal Maturity Index Modelled Arr∑TTI	Thermal Maturity Index Modelled Log₁₀ ∑TTI	Model Calculated Vitrinite Reflectance (RₒCalc)	Fraction of Kerogen Converted to Liquid Petroleum	Fraction of Kerogen Converted to Gaseous Petroleum
Multiple Burial Intervals Modelled		Time (b.p.)	Burial and thermal history of deepest horizon (Top FM19)						Deposition of FM19 begins	Oil Window	Gas Window
Surface Interval	0	150	Deposition of FM19 begins			21.0					
End Interval 1	3601	145	720.2	0.035	25.21	147.0	1.12E-01	-0.95	0.81	0.1057	0.0011
End Interval 2	3601	125	0.0	0.045	1.80	183.0	2.52E+02	2.40	1.67	1	0.9195
End Interval 3	5348	103	79.4	0.045	3.57	261.7	8.15E+05	5.91	3.60	1	1
End Interval 4	5548	100	66.7	0.030	-24.74	187.4	9.32E+05	5.97	3.64	1	1
End Interval 5	5348	97	-66.7	0.025	-10.91	154.7	9.32E+05	5.97	3.64	1	1
End Interval 6	6098	91	125.0	0.025	3.13	173.5	9.33E+05	5.97	3.64	1	1
End Interval 7	7115	70	48.4	0.042	6.97	319.8	6.37E+07	7.80	4.43	1	1
End Interval 8	7315	67	66.7	0.042	2.80	328.2	2.05E+08	8.31	4.69	1	1
End Interval 9	7115	65	-100.0	0.035	-29.10	270.0	2.34E+08	8.37	4.69	1	1
End Interval 10	7388	60	54.6	0.025	-12.87	205.7	2.34E+08	8.37	4.69	1	1
End Interval 11	8643	55	251.0	0.040	32.20	366.7	6.36E+08	8.80	4.69	1	1
End Interval 12	9198	36	29.2	0.040	1.17	388.9	3.64E+10	10.56	4.69	1	1
End Interval 13	9398	35	200.0	0.040	8.00	396.9	4.08E+10	10.61	4.69	1	1
End Interval 14	9198	34	-200.0	0.038	-26.40	370.5	4.36E+10	10.64	4.69	1	1
End Interval 15	9438	22	20.0	0.034	-2.39	341.9	4.96E+10	10.70	4.69	1	1
End Interval 16	9998	18	140.0	0.034	4.76	360.9	5.10E+10	10.71	4.69	1	1
End Interval 17	10298	16	150.0	0.034	5.10	371.1	5.26E+10	10.72	4.69	1	1
End Interval 18	9998	11	-60.0	0.034	-2.04	360.9	5.69E+10	10.75	4.69	1	1
End Interval 19	10398	0 Today	36.4	0.034	1.24	374.5	6.74E+10	10.83	4.69	1	1

(Vertical label between columns: Burial of FM19 Progresses over time)

of calculating temperature at depth over time for 15–20 horizons and display that graphically in a form similar to Fig. 5.4.

Step 2: Use the temperature and time data for each horizon to calculate the $\sum TTI_{ARR}$ ($E = 218$; $\log_e A = 61.56$) value from which calculated vitrinite reflectance can be derived for the organic-rich formations of interest using the available correlations (Fig. 5.5). These Ro_{calc} values need to be verified with at least some Ro_{meas} (or other reliable and quantified thermal maturity indicators) to validate the assumptions made in the burial and thermal history reconstruction.

Step 3: Conduct detailed petroleum transformation analysis for the specific kerogen type(s) present in the organic-rich formations of interest. This will typically require kerogen kinetic information (E and A values) for that particular kerogen or kerogen mix that may be different from the E and A values used for basin wide thermal maturity analysis in step 2. These petroleum transformation calculations are discussed in subsequent sections.

A key output from a multi-dimensional burial history is the depth and thermal maturity profile for all the formations considered at the present day. This is shown for the burial-history example described (Fig. 5.4, Table 5.1) in Table 5.2. This is essentially the depth profile penetrated by the wellbore (or reconstruction of a pseudo-wellbore) with the cumulative thermal maturation data displayed. It is for this profile that the Ro_{calc} values and Ro_{meas} values are typically compared and the petroleum transformation levels of organic-rich zones of interest are displayed.

The closer the correlation between Ro_{calc} values and Ro_{meas} values in the present-day depth profile (and forecast sub-surface temperatures versus measured subsurface

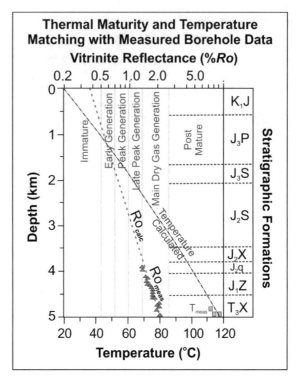

Fig. 5.5 Thermal maturity modelling results for well B4 in the Yuanba area, north-east Sichuan Basin, China (modified after Yang et al. 2016) showing a close match between measured and modelled Ro values for the lower Jurassic and Upper Triassic source rocks in the gas window

Table 5.2 Calculated Ro, $\sum TTI_{ARR}$ and petroleum transformation values at the present day for all horizons modelled in Table 5.1 and displayed in Fig. 5.4. It is this data that is particularly useful for the petroleum exploitation potential of shales as unconventional reservoirs (modified from Wood 2018a)

Thermal Maturity and Temperature Profile Today for a Single Depth Profile for an Example Deep Rift Basin										
Present-day depth profile of each formation top modelled					One-Dimensional Thermal Maturity of Present-day Formation Sequence as it Exists Today					
Formation Name	Depth to Base FM Formation (metres)	Age of Formation Top (m.y.)			Present-day Temperature (°C)	Thermal Maturity Index Modelled Arr∑TTI	Thermal Maturity Index Modelled Log₁₀ ∑TTI	Model Calculated Vitrinite Reflectance (R₀Calc)	Fraction of Kerogen Converted to Liquid Petroleum	Fraction of Kerogen Converted to Gaseous Petroleum
FM1	400	11		Near Surface	34.6	1.52E-10	-9.82	0.22	0	0
FM2	100	16	Eroded section (-300m)		24.4	Eroded	Eroded	Eroded	Eroded	Eroded
FM3	400	18			34.6	1.99E-10	-9.70	0.23	0	0
FM4	960	22			53.6	3.27E-08	-7.49	0.33	0	0
FM5	1200	34			61.8	2.44E-07	-6.61	0.37	0	0
FM6	1000	35	Eroded section (-200m)		55.0	4.61E-08	-7.34	0.33	0	0
FM7	1200	36			61.8	2.44E-07	-6.61	0.37	0	0
FM8	1755	55			80.7	1.78E-05	-4.75	0.46	0	0
FM9	3010	60			123.3	6.55E-02	-1.18	0.78	0.0634	0
FM10	3283	65			132.6	3.12E-01	-0.51	0.88	0.2681	0
FM11	3083	67	Eroded section (-200m)		125.8	1.00E-01	-1.00	0.81	0.0953	0.0010
FM12	3283	70			132.6	3.12E-01	-0.51	0.88	0.2681	0.0031
FM13	4300	91			167.2	6.04E+01	1.78	1.45	1	0.4534
FM14	5050	97			192.7	1.86E+03	3.27	2.05	1	1
FM15	4850	100	Eroded section (-200m)		185.9	7.71E+02	2.89	1.87	1	1
FM16	5050	103			192.7	1.86E+03	3.27	2.05	1	1
FM17	6797	125			252.1	1.81E+06	6.26	3.80	1	1
FM18	6797	145		Basin	252.1	1.81E+06	6.26	3.80	1	1
FM19	10398	150		Floor	374.5	6.74E+10	10.83	4.69	1	1

temperatures), the more confidence there can be in the accuracy of the assumptions made in the burial and thermal maturation model and reconstruction. Figure 5.5 shows the type of display that generates such confidence for thermal maturity and temperature matching of calculated and measured data with borehole depths conducted for a wellbore in the Yuanba region in the north-east of the Sichuan Basin, China (Yang et al. 2017).

Having constructed burial and thermal histories at several discrete locations across a basin (at boreholes or pseudo-boreholes), and satisfactorily matched the calculated and measured thermal maturity data for those points, the validated model can be extended across an entire basin. To do this requires depth maps for the key horizons across the area of interest, supported by geothermal gradient maps (for different geological time intervals). With such data, and a rigorously calculated thermal maturation model, it is possible to generate depth maps displaying temperature, thermal maturity ($\sum TTI_{ARR}$ and Ro_{calc}) and petroleum transformation for the present day and for selected past geological times.

Quantitative thermal maturity maps, on a basin-wide scale, are able to provide insight to the extent and timing of thermal maturity levels achieved by organic-rich formations present in the geological column, leading to the identification of petroleum sweet spots within the basin. Moreover, the gross rock volume of the organic-rich formations mapped to be within certain ranges of thermal maturity (e.g., the oil window or the gas window), and adjusted by the net fractions of the high-quality kerogen containing zones (e.g. applying TOC and/or HI cutoffs), can be used to quantify the volumes of petroleum present in-place, and potentially recoverable, for the basin as a whole. This three-dimensional thermal maturity mapping approach is applicable to delineating unconventional resource plays and identifying petroleum-generation kitchen areas for conventional oil and gas exploration.

5.6 Optimizers to Model Erosion and Geothermal Gradients to Match Measured Ro Profiles

Many burial histories are complex and extend over long periods of geological time with periods of erosion and variable geothermal gradients. The problem with unconformities and periods of erosion is that they represent gaps in the burial history that require estimates of the thickness of sediment originally deposited, plus the rate of deposition and rate of erosion to remove that sequence. Moreover, the amount of subsequently removed sediment may vary significantly across a basin. In some cases, differential compaction studies above and below an unconformity can provide insight to the approximate thickness of removed sedimentary sequences. Often analyst use trial and error to establish the most realistic erosion thickness and paleo-geothermal gradients for best matching predicted thermal maturity with direct measurements of thermal maturity through the entire geologic section to be modelled.

Wood (2018a) demonstrated that by combining the $\sum TTI_{ARR}$ calculated thermal maturity index with an optimizer improves the modelling precision for complex burial histories and establishes meaningful range limits on the thicknesses of sections eroded in the geological past and paleo-geothermal gradients. Setting up an optimizer to minimize the mean square error between Ro_{calc} values and Ro_{meas} values was shown to be a convenient way to find optimal solutions (accurate to about two decimal places) applying various constraints and limits with respect to specific geological formations. To achieve the most insight from such optimized thermal maturity models it is often necessary to apply constraints (e.g., maximum and minimum cumulative thicknesses or geothermal gradients allowed) across groups of formations rather than individual formations. Combining the $\sum TTI_{ARR}$ calculated thermal maturity index with an optimizer facilitates rapid and transparent basin-wide thermal maturity analysis that can incorporate pseudo-wells and cross sections across the most prospective portions of a sedimentary basin.

5.7 Fractional Transformation of Kerogen into Petroleum Quantified in Terms of the Cumulative Arrhenius Time Temperature Index $\sum TTI_{ARR}$

The Arrhenius equation is frequently used to calculate the extent to which a reaction proceeded after a certain time has elapsed knowing the first order reaction rate k. This is achieved by simply expressing the Arrhenius equation in the form of Eq. (5.8):

$$X_t = X_o e^{-kt} \tag{5.8}$$

where

X_o is the quantity of the reactant present before the reaction began;
X_t is the quantity of the reactant yet to be transformed at time t; and
X_t/X_o is the fraction of the reactant (on a scale of zero to one) yet to be converted;
k is the Arrhenius equation first-order reaction rate which is dependent on the E and A value for the specific reaction.

For kerogen, kt in Eq. (5.8) is proportional to the thermal maturation process. As temperature varies over time, the reactions involved lead to its progressive transformation into petroleum, depending upon the E and A values of the particular kerogen and its burial/heating history. It is therefore reasonable to substitute $\sum TTI_{ARR}$ Eq. (5.4) for kt in Eq. (5.8) (Wood 1988, 2017) leading to the relationship expressed as Eq. 5.9:

$$X_t/X_o \approx e^{-\sum TTIARR} \tag{5.9}$$

The approximation in these relationships recognizes that there are likely several reactions involved and the time temperature index is representing the average of those reactions.

The transformation can be normalized to a scale of 0–1 by setting the total quantity of reactant available, $X_o = 1$. The fraction of petroleum yet to be transformed from the kerogen, X_t, is then calculated by Eq. (5.10).

$$X_t = e^{-\sum TTIARR} \tag{5.10}$$

To quantifying kerogen conversion to petroleum for modelling purposes, it is typical to calculate the petroleum transformation factor (TF_t) for the average of the first order reactions involved in petroleum generation from kerogen, such that $TF_t = 1 - X_t$, meaning that TF_t can be calculated with Eq. (5.11):

$$TF_t(\text{oil}) = 1 - e^{-\sum TTIARR} \tag{5.11}$$

When $TF_t = 0$ the kerogen is thermally immature and has not yet generated any petroleum. When $TF_t = 1$, the kerogen is thermally post-mature and has generated all the petroleum it is capable of generating. It is worth noting that the condition $TF_t = 1$ does not necessarily mean that a kerogen ceases to contain any petroleum as some of the petroleum generated may remain trapped within its micropore space for sometime after the condition $TF_t = 1$ has been reached and the temperature threshold to achieve that condition has been passed. The distinction between petroleum generation and expulsion from kerogen is therefore an important one and is not easy to quantify.

One determining factor in selecting the kerogen kinetic values of $E = 218$ kJ/mol; $\log_e A = 61.56$/m.y. for the $\sum TTI_{ARR}$ thermal maturity index is that TF_t Eq. (5.11) advances from 0 to 1 for the Ro_{calc} range 0.5–1.1% (Wood 1988), i.e., it matches and models primary oil generation across the peak oil generation window as defined by the vitrinite reflectance scale. As TF_t modelled with Eq. (5.11) essential models the transformation from kerogen to petroleum across the oil generation window, it is designated TF_t (oil).

Although such a calculation is essential for the generation of liquid petroleum it provides no insight to the generation of natural gas and natural gas liquids (NGL) across a much wider temperature window (Fig. 5.1), broadly defined by an observed vitrinite reflectance range of ~0.8 < Ro < ~2.0. Natural gas and NGL generation within kerogen (and after primary migration of petroleum out of the kerogen) is complex and likely related to a number of first-order and higher-order reactions (e.g. cracking and reforming of liquid petroleum in the kerogen and in the pores and fractures of the shale). The combination of these multiple and distinct reaction types is not readily constrained or modelled by the average Arrhenius reaction kinetics selected from the E-A trend identified in Fig. 5.3. An approximation of natural gas/NGL transformation can though be obtained by applying a shift to the $\sum TTI_{ARR}$ scale such that the transformation fraction as defined by Eq. (5.11) matches gas-window thermal maturity levels, as defined in terms of the vitrinite reflectance range

~0.8 < Ro < ~2.0). By dividing the $\sum TTI_{ARR}$ scale (derived with E = 218 kJ/mol; $\log_e A$ = 61.56/m.y.) by 100 (Wood 2018b), its transformation faction, TF_t (gas) is shifted to cover the gas-generation window (defined as ~0.8 < Ro < ~2.0). This empirical shift to approximate natural gas formation (i.e., primary generation and secondary formation by cracking, reforming etc.) in relation to the modelled thermal maturity scale $\left(\sum TTI_{ARR}\right)$ is expressed as Eq. (5.12).

$$TF_t(\text{gas}) = 1 - e^{-\sum TTIARR/100} \qquad (5.12)$$

If a different pair or group of E-A kinetic values are used to calculate the $\sum TTI_{ARR}$ scale, then the denominator required to empirically calibrate TF_t(gas) to the gas-generation window (defined as ~0.8 < Ro < ~2.0) using Eq. (5.12) is likely to change. Shifting the $\sum TTI_{ARR}$ index up or down by various adjustment factors reveals the impacts on kerogen TF_t relative to the Ro scale. This simulates the impact of slower or faster reaction kinetics than E = 218 kJ/mol and $\log_e A$ = 61.56/m.y. applied to develop the standard $\sum TTI_{ARR}$ thermal maturity scale. Applying different kerogen kinetics to calculate Eqs. (5.11) and (5.12) is evaluated further in subsequent sections of this chapter.

5.8 Range of Kerogen Kinetics Observed in Shales Worldwide

Shale characterization including its petroleum-generation potential is widely applied drawing on routine geochemical analysis and pyrolysis tests (particularly Rock-Eval metrics). Despite this, in recent years, its objectives have broadened beyond considering shales source-rock capabilities to feed oil and gas into conventional porous and permeable reservoir zones. Although the basic petroleum geochemistry techniques (McCarthy et al. 2011) and analysis of petroleum potential based on the fundamental principles remains applicable, an additional focus today is to evaluate shales as standalone (self-contained), unconventional petroleum systems. This shift has also increased the significance of understanding the specific reaction kinetics of the kerogens in shales in better characterizing their petroleum potential (Wood 2017).

Figure 5.6 displays kerogen kinetics for a range of kerogen types published in term of the A and E values. The linear trend, also displayed in Fig. 5.3, (Wood 1988) was based on published pyrolysis data available at that time include analysis from the Woodford shale and Phosphoria Retort Shale from the U.S.A. (Lewan 1985) and type I, II and III kerogens from around the world (Tissot and Espitalié 1975; Tissot and Welte 1978). The trend defined by Ungerer (1990) is based on numerous samples of type I, II and III kerogens from around the world analysed by Institut Francais du Petrole up to that time (originally plotted on the laboratory scale for A expressed in/seconds units). The data of Peters et al. (2015) is for two shale

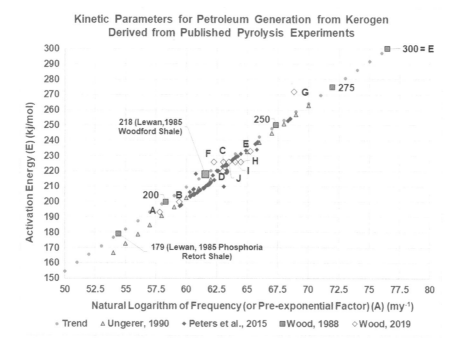

Fig. 5.6 Arrhenius equation reaction kinetics for a range of kerogens compiled from published data in terms of activation energies (E) versus the natural logarithms of the pre-exponential (frequency) factors expressed on a per-million-year time scale ($Log_e A$). The "trend" is taken from Wood (1988) and already described in Fig. 5.3. The sources of data and their locations are described in the text

samples (Kimmeridge Clay with type II kerogen from the U.K.; Monterey Shale a type IIS—sulphur rich-from California, U.S.A.) analysed at different heating ramps in detailed pyrolysis analysis. The Monterey shale samples ranges between E = 197 kJ/mol to E = 234 (mean E = 210) and the Kimmeridge clay samples range from E = 206 to E = 254 (mean E = 228 kJ/mol) with all samples falling close to the Wood (1988), Ungerer (1990) trends in Fig. 5.6.

The samples labelled A–J in Fig. 5.6 are published samples from various locations and sources compiled by Wood (2019) because they included S2-peak pyrograms measured for at least three different heating rates (making kinetic modelling possible). These samples are:

A. Kerogen from Pingliang Formation (PL-M-O2p, marine shale, Middle Ordovician outcrop sample from the Erdos/Ordos Basin, China (Liao et al. 2018).
B. Kerogen from Yangchang Formation (YC-L, T3y, lacustrine shale, Upper Triassic borehole Zheng-8 in the Erdos/Ordos Basin, China (Liao et al. 2018).
C. Kerogen from AP22 Green River Formation (Reynolds and Burnham 1995).
D. Kerogen from the Kimmeridge Clay, Draupne Formation, Late Jurassic, Northern North Sea, Norway (Reynolds and Burnham 1995).

E.   Kerogen from the Phosphoria Shale, Early Permian, Retort Mountain Quarry in Beaverhead, Montana, U.S.A. (Reynolds and Burnham 1995).

F.   South Elwood-composite Monterey Formation, Miocene shale sample (13.58 wt% S) borehole Santa Barbara-Ventura Basin, California, U.S.A. (Reynolds et al. 1995).

G.   Naples Beach Monterey Formation, Miocene outcrop sample (8.88 wt% S) Santa Barbara County, California, U.S.A. (Reynolds et al. 1995).

H.   Los Angeles Basin Modelo, Middle Miocene shale outcrop sample (6.16 wt% S) Bel Air, California, U.S.A. (Reynolds et al. 1995).

I.   Montney Triassic shale sample 172208 (immature, Ro = 0.59 wt%) Western Canada Sedimentary Basin, Alberta, Canada (Romero-Sarmiento et al. 2016).

J.   Doig Triassic Shale sample 173182 (early mature, Ro = 0.71 wt%) Western Canada Sedimentary Basin, Alberta, Canada (Romero-Sarmiento et al. 2016).

What is significant about all the samples plotted in Fig. 5.6 is that they are from a number of petroleum-rich basins around the world, analysis in several different laboratories by different research groups spread over a period of some five decades. Accepting that there will be a certain amount of analytical error involved the distribution of these samples, what is striking is that they are all distributed reasonably close to the trend defined by Wood (1988) and that the kinetic point of E = 218 kJ/mol and $\log_e A = 61.56$ used for the $\sum TTI_{ARR}$ thermal maturity index is located at a reasonably central point in that distribution of published samples. Further work compiling much more kerogen kinetic data from around the world will undoubtedly refine the trend of kerogen kinetics, but for the purposes of petroleum generation and thermal maturity analysis for shales that trend provides a meaningful basis from which to conduct meaningful quantitative analysis.

What this data also displays is that it is not so much the type of kerogen (e.g., type I, II, III, IV) that determines at what temperature and time a shale will generate its petroleum, but rather the average reaction kinetics of the collective kerogen constituting the organic content of the shales. It is therefore paramount that kerogen kinetics is determined as accurately as possible for shales to evaluate their potential contributions as conventional source rocks and as unconventional reservoirs.

5.9   Multi-heating-rate Pyrolysis S2 Peaks to Determine Kerogen Kinetics Distributions

Reaction kinetic distributions for kerogens are typically expressed in terms of activation energies (E) and pre-exponential factors (A). These E-A values are determined from pyrolysis measurements (specifically targeting the S2-pyrolysis peak) performed at three or more different rates of heating (or heating ramps). A widely-used method find the range of E values at a fixed A value (expressed in /s or /min on a laboratory time scale) that best fits the suite of S2 pyrolysis peaks obtained at different heating rates (e.g., Peters et al. 2015). The validity of this fixed-A method has been questioned by Wood (2019), because its arbitrary assumption (fixed A) leads to a spread of E values in the E distribution for each specific kerogen analysis that does not follow the average E-A trend of kerogen reactions observed for kerogens in general (Fig. 5.6). The kinetic distributions obtained by using variable E and variable A values, involve more complex fitting algorithms, but display good fits involving ranges of kinetic (E-A) values in the E-A distribution of specific kerogens that do follow the more realistic average E-A kerogen kinetics trend defined for several decades (Wood 2019).

Accurately fitting multi-rate S2-pyrolysis peaks (normalized on a scale 0–1) and their associated transformation curves (cumulative scale of 0–1) can be readily achieved with the aid of an optimizer using Eq. (5.4) ($\sum TTI_{ARR}$ to establish the reaction increments) and Eq. (5.11) (*TFt* to establish the associated transformation fractions) calculated with a range of E-A combinations. Such fitting is effective when conducted at 1 °C temperature intervals over the reaction temperature range of 250–700 °C. The methodology for fitting multi-heating-rate pyrolysis data in this way (Wood 2019) requires two steps. The first step uses the optimizer to locate a specific E-A kinetic pair that most accurately matches the S2 peak temperatures (not T_{max}) of the three or more for multi-rate pyrolysis curves being fitted. The second step establishes the E-A kinetic pair identified from as its modal focus for more detailed S2 curve and/or transformation curve fitting. The optimizer establishes a distribution of up to eleven distinct reaction kinetics (E-A values) that best fits the curves and is free to select whatever E-A combinations best achieve that objective (i.e., it is not constrained to a single A value as used by the traditional method).

Figure 5.7 illustrates an example of S2 peak and transformation fraction curve fits to the Montney Triassic shale sample 172208 (immature, Ro = 0.59 wt%) Western Canada Sedimentary Basin, Alberta, Canada (Romero-Sarmiento et al. 2016), already identified as sample I in Fig. 5.6, using the variable E-A fitting method described. Note that the best-fit solutions almost perfectly overlies the digitized data for these curves (the optimizer achieved that fit with a mean square error of 6.39E−01), as it does also for heating rates of 5 and 25 °C/min (Wood 2019). The weighted-average of the reaction kinetics distribution for the best fit displayed is: E = 227.76 kJ/mol; $Log_e A = 64.23$/m.y., which are close to the kinetic values published for that sample (i.e., E = 226; $log_e A = 64$; Romero-Sarmiento et al. 2016).

The kerogen kinetic distribution that achieves the best fit displayed in Fig. 5.7 is listed in the left side of Table 5.3. Also shown for comparison in Table 5.3 (right side) is the best-fit kerogen kinetics distribution using a fixed A value. Although, in this case, the weighted average of the kinetic distributions calculated by the two different methodologies are similar that is not always the case (e.g., compare Tables 4 and 5 in Wood 2019). Moreover, when these two complete kinetic distributions are applied to specific burial and heating histories for a shale formation, they will lead to petroleum transformation occurring to different degrees at different times before present.

Fitting multi-heating-rate pyrolysis curves (at three or more different heating rates) in the way described here does lead to confidence in the kerogen kinetics, with error limits that depend on the mean squared error associated with the best fit. Typically, there is a narrow range of kerogen kinetic distributions that will provide

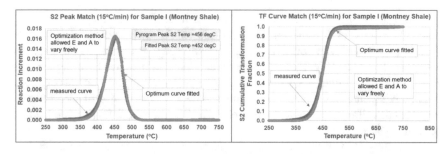

Fig. 5.7 Reaction increments (left) and transformation factors (right) for Montney Shale sample I digitised from Romero-Sarmiento et al. (2016) for a heating rate of 15 °C/min and fitted with the optimization method proposed by Wood (2019)

Table 5.3 E-A reaction kinetic distributions that produced optimum fits to the S2-peak shapes for Montney Shale sample I at three different heating rates. The left-side distribution optimizer was allowed to vary E and A, for right side distribution the optimizer was constrained to solutions using a fixed A value that matched its S2 peak temperature. The fitted curves for one heating rate (15 °C/min) using the left-side distribution is illustrated in Fig. 5.7

Montney Shale Sample Kerogen Kinetics Obtained from Pyrolysis Curve Fitting						
	Multi-rate Pyrolysis Curves fitted with Variable E-A Distribution			**Multi-rate Pyrolysis Curves fitted with Variable E- Fixed A Distribution**		
Eleven Kerogen Kinetics Achieving Best Fit	**E (kj/mol)**	**Log$_e$A (/my)**	**Fraction Selected (f)**	**E (kj/mol)**	**Log$_e$A (/my)**	**Fraction Selected (f)**
KC#1	208.39	58.99	0.0008	214.97	64.54	0.0659
KC#2	222.00	63.24	0.3500	220.17	64.54	0.0012
KC#3	227.43	64.11	0.1286	226.91	64.54	0.3312
KC#4	228.01	64.44	0.3542	229.00	64.54	0.3500
KC#5	228.13	62.74	0.0570	229.81	64.54	0.0061
KC#6	234.61	69.02	0.0214	231.72	64.54	0.0013
KC#7	237.28	66.41	0.0710	232.02	64.54	0.0005
KC#8	241.73	61.04	0.0008	232.26	64.54	0.0982
KC#9	267.50	65.45	0.0029	235.17	64.54	0.0672
KC#10	281.42	67.56	0.0035	239.22	64.54	0.0633
KC#11	312.59	75.57	0.0098	250.65	64.54	0.0150
		Sum (F):	1.0000		Sum (F):	1.0000
Mode of Distribution:	228.01	64.44		229.00	64.54	
Weighted Average of Distribution:	**227.76**	**64.23**		**229.09**	**64.54**	
MSE Fit Error:		0.6391			0.6576	

good fits to the multi-heating-rate pyrolysis curve data. This is found not to be the case for single-heating rate pyrolysis curves for which there are multiple possible kinetic distributions that can be found to adequately fit them (Wood 2019). The claim by Waples (2016) that adequate kerogen kinetics can be derived from single-heating-rate pyrolysis data is therefore not supported by detailed analysis of such data with optimizers.

What pyrolysis-S2-peak fitting reveals is that very few kinetic reactions distributions are able to accurately (e.g. to the level illustrated by Fig. 5.7) fit a set of S2 peaks related to three or more multi-rate pyrolysis analyses of a single sample. However, if E-A values are able to vary in an unconstrained way when fitting a single S2 curve, multiple potential solutions can be found. It is therefore untenable to attempt to invert accurate reaction kinetics distributions from single-heating-rate pyrolysis data. Based on the E-A distributions measured for kerogens displayed in Fig. 5.6, the closer the best-fit, multi-heating-rate kinetic distribution lies to the established E-A trend, the more confidence can be placed on the likelihood of those kinetic representing the range of reactions likely to prevail during the thermal maturation of a specific shale.

Whereas it is considered feasible to model the thermal maturity level of a shale, linked to the vitrinite reflectance scale, using a single set of kerogen kinetics (i.e., E $= 218$ kJ/mol; $\log_e A = 61.56$/m.y.), what the shapes of typical pyrolysis S2 peaks of shales reveal is that they are likely made up of a distribution of reactions (i.e., several E-A values). That set of reactions, collectively, will determine the transformation rates and extents at specified temperatures of a particular shale into petroleum. In a shale with a homogeneous single kerogen type it is realistic to expect a suite of first-order chemical reactions to be contributing to the formation of petroleum (a complex mixture of many individual hydrocarbon molecules) and in there configuration of the various chemical bonds (e.g., C–C, C=C, C–H, C–S) present in the hydrocarbon molecules contained within kerogen as it thermally matures. For a shale made up of two or more kerogen types in significant proportions, the number of reactions and the kinetic distributions that can fit multi-heating-rate pyrolysis S2 peaks are likely to be more complex, reflecting the greater number of hydrocarbon molecules and reactions likely to be involved.

5.10   Kinetics for Shales with Mixed Kerogen Types and Their Impact on Petroleum Transformation

Applying a range of E-A values/distributions to the Arrhenius equation it is possible to model the time-temperature progress (i.e. linked to specific burial and thermal histories) of petroleum transformation reflecting realistic kinetics for individual kerogen types or mixtures of two or more kerogen types. Wood (2018b) showed that such modelling could be informative if compared at both geological (/millions of years) and laboratory (/s or /min) time scales. Individual kerogens and mixed kerogens produce distinctive simulated S2 peak and cumulative transformation curve shapes which depend on the E-A values involved. Detailed analysis of the simulated curves produced identify that certain transformation curve characteristics (e.g., maximum transformation gradient and the temperature at which that maximum gradient on a 1 °C interval basis occurs, average transformation gradient between 10 and 90% conversion, and the temperature spread from 10 to 90% conversion) can readily discriminate between the transformation curves of shales with single reactions kinetics (one E-A pair), and mixtures of reaction kinetics (two or more E-A pairs). Furthermore, these metrics can discriminate between transformation curves at geological and laboratory scales. Figure 5.8 shows how some of these metrics when displayed in cross plots can readily distinguish reaction mixtures from single reactions at the geological time scales and at the faster heating rates at the laboratory scales.

The kerogen reactions used for the analysis in Fig. 5.8 come from the E-A trend defined in Fig. 5.3. Just the E values in kJ/mol are displayed. The mixture involving three reactions (those in the centre of the upper graph in Fig. 5.8) are mixed in the proportions 20%:60%:20%. For example, the three-reaction mixture labelled E180–E200 means a mixture of 20% E180:60% E190:20% E200. Such relationships could be used at the laboratory scale to potentially distinguish mixed kerogens (generally broader S2 peaks and flatter transformation curves) from single kerogens (generally narrower pyrolysis S2 peaks and steeper cumulative transformation curves). These metrics therefore offer the potential to more-precisely define the petroleum generation and transformation progress relative to time and temperature for shales with complex mixtures of kerogens.

This type of analysis can help to distinguish zones within shales or burial depth ranges ith high potential for exploitation (i.e., sweet spots) as unconventional reservoirs. Once the kerogen kinetics (single reaction or mixed) is identified, the maximum transformation gradient and its associated temperature can be calculated. This information can then be used to enable exploitation to be focused around a narrow (present-day or paleo-) temperature range and/or narrow depth-of-burial range on a geological scale. Such depth and temperature intervals will vary according to the kerogen kinetics, the burial history of the shale formation and the geothermal gradients impacting the shale over time. The detailed characterization of petroleum transformation curves from the S2 peaks generated by pyrolysis tests of shales and/or kerogens with E-A values expressed at the laboratory time scale can also help to define the most useful metrics to seek from pyrolysis tests to assist in inverting laboratory scale analysis into a geological scale exploration/exploitation tool.

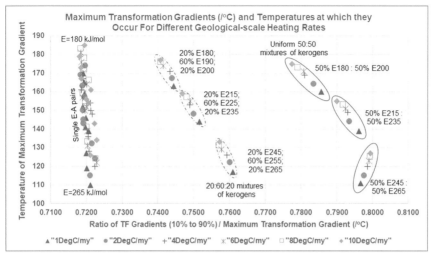

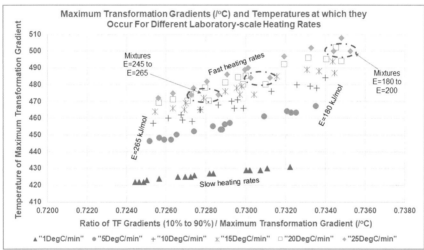

Fig. 5.8 Maximum transformation temperature gradient for S2 peak versus the ratio of the average transformation gradient between 10% and 90% transformation (on a per °C scale) to the maximum transformation gradient (/°C). These metrics are displayed for a wide range of heating rates, kerogen kinetics and reaction mixtures at the geological time scale (upper graph) and the laboratory time scale (lower graph) (from Wood 2018b)

5.11 Possible Non-kinetic Contributions to the S2 Peaks from Highly Mature Shales and the Influence of Evolving Kerogen Porosity

Most kinetic information extracted from multi-rate-heating pyrolysis tests of shales is conducted on thermally immature, or sometime early mature shales (i.e., with vitrinite reflectance less than about 0.7%). This is because the pyrolysis curves, particularly the S2 peaks, for thermally mature shales, are much more difficult to interpret and frequently described as anomalous. Often, such thermally mature shale samples display much broader S2 peaks with flatter cumulative transformation curves.

Thermally-mature shales are likely to contain petroleum fluids, particularly natural gas that was generated at early stages of thermal maturity. Some of those fluids are likely to be trapped in kerogen nanopores. Such trapped fluids may not be released by the S0 and S1 heating ramps, but only released from the sample as gas at a temperature range within the S2 pyrolysis heating ramp at which gas expands sufficiently to break out of those nanopores. Such a process is clearly not a first-order chemical reaction, but a change to the physical structure of the shale, induced by gas under increasing pressure, once certain temperatures are reached. This could explain the broader S2 peaks often displayed by thermally mature shale samples. If this interpretation is correct then it could be used to identify potential shale gas potential in thermally mature shale zones.

Kerogen in shales is known to have a complex microstructure (Yang et al. 2016) that evolves and further develops as thermal maturity levels increase. Studies of thermally-mature shales from several areas have identified increased nanopores (<1 nm diameter), together with a more expanded network of micropores (<2 nm) and mesopores (50 nm) compared to immature samples (Chalmers et al. 2009, 2012; Clarkson et al. 2013; Wood and Hazra 2017).

Processes, affecting the structural fabric of shales as well as second-order reactions, seem likely to be influencing the pyrolysis S2 peak characteristics of thermally mature shales, in addition to first-order chemical reactions associated with the transformation of kerogen into petroleum fluids. This is particularly so for shales that has remained at peak thermal maturity for many tens of millions of years (e.g., some of the Permian Raniganj shales of India, Wood and Hazra 2018). Figure 5.9 expresses diagrammatically how the chemical reaction and porosity development process may work systematically to determine the evolving shape of the pyrolysis S2 peak of at least some shales as their thermal maturity advances (as illustrated by the bottom-right diagrammatic pyrolysis S2 peak in Fig. 5.9).

Thermally immature shales samples are likely to display S2 pyrolysis peaks more closely linked to first-order reactions kinetics expressed by Arrhenius equation functions (e.g. $\sum TTI_{ARR}$). For a single type of kerogen those reactions are more likely to be linked to a single or narrow range of E-A reaction kinetics and, consequently, the pyrolysis S2 peaks they generate are more likely to be narrow and symmetrical in terms of their temperature distributions. For thermally-immature shales with mixed kerogen types the pyrolysis S2 peaks they generate are more likely to be broader and

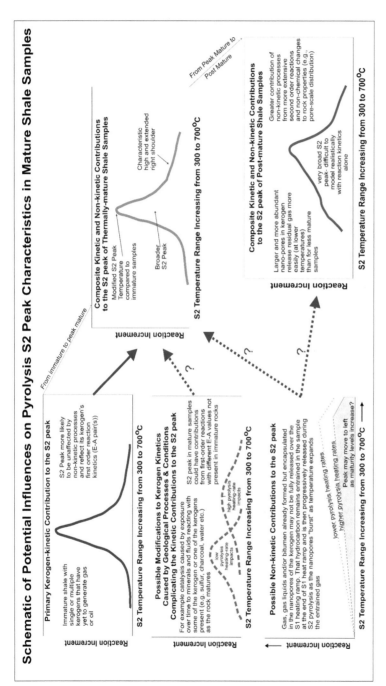

Fig. 5.9 Diagrammatic representation of how the pyrolysis S2 peak shapes of shales might evolve from thermally immature states to post-mature states highlighting the roles of first-order reaction kinetics and other second-order reactions and non-kinetic processes

asymmetrical. If the mixed kerogens have very distinctive E-A kinetic values, and at least two are both present in significant concentrations, the pyrolysis S2 peaks can be bimodal or even multi-modal in terms of their temperature distributions. If there is one dominant kerogen type in a shale but others are present in small but significant fractions then either the left shoulder or the right shoulder of the S2 peak can become larger depending upon the relative E-A values of the main and subordinate kerogens present in the shale. Wood and Hazra (2018) interpreted the slightly exaggerated right-side shoulder of pyrolysis S2 peaks of thermally mature Permian shales from the Raniganj basin in these terms (as illustrated by the top-right diagrammatic S2 peak in Fig. 5.9).

As thermal maturity advances it is also possible that other components within a shale, or its kerogens, might influence the reaction kinetics of petroleum generation, either to accelerate or decelerate the reactions. Lewan (1985) showed that sulphur present within some kerogen changes its reaction kinetics. Older shales exposed to high-temperature fluids transiting the formation for millions of years have the potential to bring the kerogen into contact with various trace elements and metals leading to catalytic effects on their kerogen reaction kinetics. Some shales (e.g., Raniganj shales, India) contain significant quantities of inertinite (Type IV kerogen made up primarily of fossilized charcoal and burned wood fragments). The charcoal, in particular can in certain conditions act as catalyst support carbon (Larson and Walton 1940; Juntgen 1986; Trogadas et al. 2014) and thereby accelerate the reaction kinetics of some Type III or other kerogen it comes into contact with.

References

Arrhenius S (1889) Über die Reaktionsgeschwindigkeit bei der Inversion von Rohrzucker durch Säuren. Z Phys Chem 4:226–248

Chalmers G, Bustin RM, Power I (2009) A pore by any other name would be as small: the importance of meso- and microporosity in shale gas capacity. In: AAPG annual convention and exhibition, Denver, Colorado, p 1

Chalmers GR, Bustin RM, Power IM (2012) Characterization of gas shale pore systems by porosimetry, pycnometry, surface area, and field emission scanning electron microscopy/transmission electron microscopy image analyses: examples from the Barnett, Woodford, Haynesville, Marcellus, and Doig units. AAPG Bull 96:1099–1119

Chen Z, Liu X, Guo Q, Jiang C, Mort A (2017) Inversion of source rock hydrocarbon generation kinetics from Rock-Eval data. Fuel 194:91–101

Clarkson CR, Solano N, Bustin RM, Bustin AMM, Chalmers GRL, He L, Melnichenko YB, Radliński AP, Blach TP (2013) Pore structure characterization of North American shale gas reservoir using USANS/SANS, gas adsorption, and mercury intrusion. Fuel 103:606–616

Cokar M, Ford B, Kallos MS, Gates ID (2013) New gas material balance to quantify biogenic gas generation rates from shallow organic-matter-rich shales. Fuel 104:443–451

Cornford C (2009) Source rocks and hydrocarbons of the North Sea, Chapter 11 In: Glennie KW (ed) Petroleum geology of the North Sea: basic concepts and recent advances, 4th edn, pp 376–462

Dieckmann V (2005) Modelling petroleum formation from heterogeneous source rocks: the influence of frequency factors on activation energy distribution and geological prediction. Mar Pet Geol 22:375–390

Donelick RA, O'Sullivan PB, Ketcham RA (2005) Apatite Fission-Track Analysis. Rev Mineral Geochem 58:49–94

Espitalié J, Laporte JL, Madec M, Marquis F, Leplat P, Pauletand J, Boutefeu A (1977) Methoderapide de caracterisation des roches meres, de leur potential petrolier et de leudegred'evolution. Inst Fr Pét 32:23–42

He S, Middleton M (2002) Heat flow and thermal maturity modelling in the Northern Carnarvon Basin, North west Shelf, Australia. Mar Pet Geol 19:1073–1088

Huntsberger TL, Lerche I (1987) Determination of paleo heat-flux from fission scar tracks in apatite. J Pet Geol 10(4):365–394. https://doi.org/10.1111/j.1747-5457.1987.tb00580.x

Jarvie DM (2014) Components and processes affecting producibility and commerciality of shale resource systems. Geologica Acta 12(4):307–325, Alago Special Publication. https://doi.org/10.1344/geologica Acta 2014.15.3

Juntgen H (1986) Activated carbon as a catalyst support: a review of new research results. Fuel 65:1436–1446

Larson EC, Walton JH (1940) Activated carbon as a catalyst in certain oxidation-reduction reactions. J Phys Chem 44(1):70–85. https://doi.org/10.1021/j150397a009

Larter S (1989) Chemical modelling of vitrinite reflectance evolution. Geol Rundsch 78:349–359

Lewan MD (1985) Evaluation of petroleum generation by hydrous pyrolysis experimentation. Philos Trans R Soc Lond Ser A 315:123–134

Liao L, Wang Y, Chen C, Shi S, Deng R (2018) Kinetic study of marine and lacustrine shale grains using Rock-Eval pyrolysis: implications to hydrocarbon generation, retention and expulsion. Mar Pet Geol 89:164–173

Lopatin NV (1971) Temperature and geologic time as factors in coalification (in Russian). Akademiya Nauk SSSR Izvestiya, Seriya Geologicheskaya 3:95–106

McCarthy KR, Niemann M, Palmowski D, Peters K, Stankiewicz A (2011) Basic petroleum geochemistry for source rock evaluation. Schlumberger Oilfield Rev (Summer 2011):32–43

Mohamed AY, Whiteman AJ, Archer SG, Bowden SA (2016) Thermal modelling of the Melut basin Sudan and South Sudan: implications for hydrocarbon generation and migration. Mar Pet Geol 77:746–762

Nielsen SB, Barth T (1991) Vitrinite reflectance: comments on "A chemical kinetic model of vitrinite maturation and reflectance" by Alan K. Burnham and Jerry J. Sweeney. Geochim Cosmochim Acta 55:639–641

Nunn JA, Sleep NH, Moore WE (1984) Thermal subsidence and generation of hydrocarbons in Michigan basin. AAPG Bull 68:296–315

Pepper AS, Corvi PJ (1995) Simple kinetic models of petroleum formation: part I—oil and gas generation from kerogen. Mar Pet Geol 12:291–319

Peters KE, Burnham AK, Walters CC (2015) Petroleum generation kinetics: single versus multiple heating-ramp open-system pyrolysis. AAPG Bull 99(4):591–616

Reynolds JG, Burnham AK (1995) Comparison of kinetic analyses of source rocks and kerogen concentrates. Org Geochem 23(1):11–19

Reynolds JG, Burnham AK, Mitchell TO (1995) Kinetic analysis of California petroleum source rocks by programmed temperature micropyrolysis. Org Geochem 23(2):109–120

Rice DD, Claypool GE (1981) Generation, accumulation, and resource potential of biogenic gas. Am Assoc Pet Geol Bull 65:5–25

Romero-Sarmiento M-F, Euzen T, Rohais S, Jiang C, Littke R (2016) Artificial thermal maturation of source rocks at different thermal maturity levels: application to the Triassic Montney and Doig formations in the Western Canada Sedimentary Basin. Org Geochem 97:148–162

Schneider F, Dubille M, Montadert L (2016) Modeling of microbial gas generation: application to the eastern Mediterranean "Biogenic Play". Geol Acta 14(4):403–417

Shurr GW, Ridgley JL (2002) Unconventional shallow biogenic gas systems. Am Assoc Pet Geol Bull 86(11):1939–1969

Stainforth JG (2009) Practical kinetic modeling of petroleum generation and expulsion. Mar Pet Geol 26:552–572

Sweeney JJ, Burnham AK (1990) Evaluation of a simple model of vitrinite reflectance based on chemical kinetics. AAPG Bull 74(10):1559–1570

Tissot BP, Espitalié J (1975) L'evolution thermique de la matiere organique des sediments: applications d'une simulation mathematizue. Revue de l'Institut Français du Petrole 30:743–778

Tissot BP, Welte DH (1978) Petroleum formation and occurrence; a new approach to oil and gas exploration. Springer-Verlag, Berlin, Heidelberg, New York

Trogadas P, Fuller TF, Strasser P (2014) Carbon as catalyst and support for electrochemical energy conversion. Carbon 75:5–42

Ungerer P (1990) State of the art of research in kinetic modelling of oil formation and expulsion. In: Durand B, Behar F (eds) Proceedings of the 14th international meeting on organic geochemistry, Paris, France, 18–22 Sept 1989. Org Geochem 16:1–25

Waples DW (1980) Time and temperature in petroleum generation and application of Lopatin's technique to petroleum exploration. Am Assoc Pet Geol Bull 64:916–926

Waples DW (2016) Petroleum generation kinetics: single versus multiple heating-ramp open-system pyrolysis. Discussion. AAPG Bull 100:683–689

Whiticar MJ (1994) Correlation of Natural gases with their sources. In Magoon J, Dow WG (eds) The petroleum system- from source to trap American Association of Petroleum Geologists, Memoir, vol 60, pp 261–283

Wood DA (1988) Relationships between thermal maturity indices of Arrhenius and Lopatin methods: implications for petroleum exploration. AAPG Bull 72:115–135

Wood DA (2017) Re-establishing the merits of thermal maturity and petroleum generation multi-dimensional modelling with an Arrhenius equation using a single activation energy. J Earth Sci 28(5):804–834. https://doi.org/10.1007/s12583-017-0735-7

Wood DA (2018a) Thermal maturity and burial history modelling of shale is enhanced by use of Arrhenius time-temperature index and memetic optimizer. Petroleum 4:25–42

Wood DA (2018b) Kerogen conversion and thermal maturity modelling of petroleum generation: integrated analysis applying relevant kerogen kinetics. Mar Pet Geol 89:313–329. https://doi.org/10.1016/j.marpetgeo.2017.10.003

Wood DA (2019) Establishing credible reaction-kinetics distributions to fit and explain multi-heating rate S2 pyrolysis peaks of kerogens and shales. Adv Geo-Energy Res 3(1):1–28. https://doi.org/10.26804/ager.2019.01.01

Wood DA, Hazra B (2017) Characterization of organic-rich shales for petroleum exploration & exploitation: a review- part 2: geochemistry, thermal maturity, isotopes and biomarkers. J Earth Sci 28(5):758–778

Wood DA, Hazra B (2018) Pyrolysis S2-peak characteristics of Raniganj shales (India) reflect complex combinations of kerogen kinetics and other processes related to different levels of thermal maturity. Adv Geo-Energy Res 2(4):343–368. https://doi.org/10.26804/ager.2018.04.01

Yang H, Zhang Y, Ma D, Wen B, Yu S, Xu Z, Qi X (2012) Integrated geophysical studies on the distribution of Quaternary biogenic gases in the Qaidam Basin, NW China. Pet Explor Dev 39(1):33–42

Yang R, He S, Li T, Yang X, Hu Q (2016) Origin of over-pressure in clastic rocks in Yuanba area, northeast Sichuan Basin, China. J Nat Gas Sci Eng 30:90–105

Yang R, He S, Hu Q, Hu D, Yi J (2017) Geochemical characteristics and origin of natural gas from Wufeng-Longmaxi shales of the Fuling gas field, Sichuan Basin (China). Int J Coal Geol 171:1–11

Chapter 6
Sedimentary Biomarkers and Their Stable Isotope Proxies in Evaluation of Shale Source and Reservoir Rocks

6.1 Introduction

The emergence of unconventional shale source and reservoir rocks as a reserve for abundant natural gas, 'the shale gas' in last decade has prompted the re-investigation of source and reservoir characteristics of the conventional petroleum (oil and gas) systems, and other organically rich and thermally mature shale formations across the globe for understanding the processes and pathways, which lead to the formation of a prolific or potential gas shale system. Amongst several properties of the source and reservoir rocks, the sedimentary organic facies abundance, quality and thermal maturity are significant, as organic-matter serves as a precursor source material that eventually, through the thermal cracking in the subsurface, produces the gaseous hydrocarbon. Organic-matter provenance, depositional settings, and appropriate thermal exposure have critical implications upon the fate and properties of shale rocks for their development into a source cum reservoir rock in a gas shale system.

The sedimentary organic-matter is composed essentially and primarily of carbon and hydrogen bearing groups with minor contribution of hetero elements such as nitrogen, sulphur and oxygen. Thus, any organic-matter characterization, basically, concerns with the forms in which the major contributor, carbon, is present in the rocks and the transformations that occur during deposition and preservation along with the sediments, leading to organically rich or lean rock. Evidently, the sedimentary rocks concentrate most of the carbon on Earth, a large part (~82%) of which is in the form of inorganic carbonates and remaining (~18%) is organic carbon. There exists a dynamic equilibrium between the oxidised (or inorganic e.g. CO_2; HCO_3^-) and reduced (or organic $C_6H_{12}O_6$; CH_4) forms of carbon, which is exhibited in the global cycling of carbon. The carbon compounds move from one carbon reservoir or sink to another and the retention, alteration and migration in similar or changed forms of original carbon in its various reservoirs occurs at short and deep geological time scale.

© Springer Nature Switzerland AG 2019
B. Hazra et al., *Evaluation of Shale Source Rocks and Reservoirs*,
Petroleum Engineering, https://doi.org/10.1007/978-3-030-13042-8_6

6.2 Organic Carbon in Shale Source Rock

The total organic carbon (TOC) content is a diagenetic product of natural biopolymers such as proteins, carbohydrates, lipids and lignin, synthesized by the living organisms. In recently buried sediments of shallow depth, the diagenetic processes result in two quantitatively different, but important organic fractions. First is the kerogen, which forms the bulk of organic-matter, and the second are the free molecules of lipids, which include the hydrocarbons and related compounds (Tissot and Welte 1984). Kerogen constitutes about 80–90% of the TOC. Various chemical and physical transformations occur in kerogen as it gets successively buried along with the sediments at different depths in the subsurface, ultimately leading to the formation of liquid and gaseous hydrocarbons, through diagenetic, catagenetic and metagenetic stages (Fig. 6.1).

About 10–20% of sedimentary organic-matter is formed of lipids synthesized naturally by the former living organisms. It comprises of carbon homologs of more than C_{15+} molecules that have recognizable biological and chemical structure and constitutes the biomarkers (Tissot and Welte 1984; Hunt 1996). These comprise the bitumen fraction in the TOC. The carbon skeleton in biomarker compounds gets only minimal or slightly altered during the digenetic processes, thus providing useful information on paleo-events of its deposition and preservation.

The kerogen and biomarkers compounds in sedimentary organofacies serve as vital proxy for the decoding the source, environment of deposition and preservation and transformation of organic-matter over time. The elemental compositions and ratios of C, H and O in kerogen are important indicators of lacustrine, marine, and terrigenous sourced organic-matter and also of the extent to which it has experienced thermal exposure and transformation. The biomarkers derived from earlier organisms carry a high degree of taxonomic specificity and allow the reconstruction of past organismic diversity (Brocks and Summons 2003). Biomarkers help in deciphering the origin, depositional settings and thermal maturity of organic-matter and are particularly used in source rock and oil correlations in petroleum systems.

The nature of origin, stability and reactivity of these two forms of organic carbon govern the long term preservation, deposition and transformation of these in the shale source rocks. These properties are quite distinguishing of the kerogen and biomarkers, and have led to the distinct techniques for their qualitative and quantitative trace measurements (Fig. 6.2).

Trace concentrations of organic carbon, specifically the biomarkers, pose a major challenge for the identification and quantification in sediments. The corroboration of the compositional concentrations with the stable isotopic ratios serve as powerful proxies for deciphering the deposition and evolution of shale source rocks.

The chapter deals in detail with the minor fraction of the TOC, 'the biomarkers', its origin and occurrence in shale source and reservoir rocks, the transformations that occur with successive burial, its isotopic signatures and the crucial information provided on the organic-matter origin, depositional environment and thermal maturity along with the contemporary analytical approaches for its characterization.

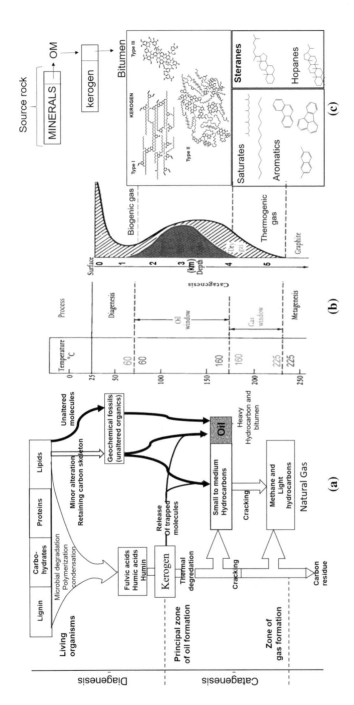

Fig. 6.1 Thermal transformation of organic-matter in the subsurface (modified after Mani 2019)

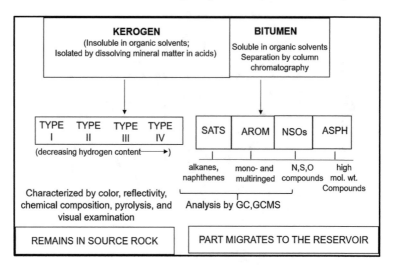

Fig. 6.2 Distinctive properties and identification techniques of kerogen and bitumen compounds

6.3 Sedimentary Biomarkers

Biological markers are very complex molecular fossils, and are also known as biomarkers or geochemical fossils. These are natural products, which are derived from biochemicals, primarily lipids, through a specific biosynthetic pathway that existed in once-living organisms. These organic compounds have high degree of taxonomic specificity with limited number of definite sources. Biomarkers encode information about the biodiversity of ancient organisms, their trophic associations, and environmental conditions of inhabitation and bear the imprints of element cycling, sediment and water chemistry, redox conditions, and temperature histories (Brocks and Summons 2003; Peters et al. 2005). Biomarkers are useful because they can provide information on organic-matter input source, environmental conditions during its deposition and burial, the thermal maturity experienced by rock or oil, the degree of biodegradation, some aspects of source rock mineralogy, lithology and age (Peters et al. 2005). The property of biomarkers which makes them a useful proxy for deciphering the Earth's paleo conditions during geological times is their recalcitrant nature against geochemical changes and easily analysability in sediments (Hallmann et al. 2011). Hydrocarbon biomarkers are stable for billions of years if they are enclosed in intact sedimentary rocks that have only suffered a mild thermal history (Hallmann et al. 2011).

6.3.1 Origin and Preservation of Biomarkers in Sediments

The sedimentary biomarkers preserve the signatures of ancient life, which manifests itself in three domains—the archaea and eubacteria (prokaryotes) and eukarya (eukaryotes or higher organisms) (Fig. 6.3). The prokaryotes consist of millions of unicellular archaeal and eubacterial species, which are distinguished mainly by their diverse biochemistry and the habitats in which they grow (Briggs and Summons 2014). The relatively simple shapes limit classification of prokaryotes based on morphology. The eukaryotes contain a nucleus which is membrane-bound and have complex organelles for specialized functions of life. Eukaryotes are classified mainly by morphology. Eukaryotic microorganisms include algae, protozoa, and fungi (moulds and yeasts). All higher multicellular organisms are also eukaryotes.

The average preservation rate of organic carbon is less than 0.1%, and is controlled by the geological and geochemical conditions of sedimentation. High organic productivity, nutrient supply which is brought by rivers and upwelling currents, low oxygenation levels of water column and sediments (less than 0.2 ml/L of water), restricted water circulation and lack of bioturbation, fine-grained sediment particles (<2 μm), and an optimum sedimentation rate favour the preservation of organic-matter in the sediments.

The accumulation and preservation of organic-matter is also controlled by the sediment grain size. Clay type sediments adhere maximum organic material, followed by the carbonates. As a result, amongst the sedimentary rocks, fine grained shales have the highest percentage of organic carbon, only seconded by the carbonate rocks (Mani 2019). The similar hydrodynamic behaviour of the organic-matter tends to get deposited preferentially with fine-grained mud. Unlike sand, fine grained mud more readily excludes oxygen-rich water below the sediment-water interface, thereby enhancing anoxia when it develops.

Post deposition, the biopolymers such as carbohydrates, proteins etc. are recycled quickly and the more resistant molecules such as lipids get concentrated in the sediments. A variety of chemical and biological processes operate on the organic-matter. Large fractions of lipids and other low-molecular weight components react via oxidation, reduction, polymerization, rearrangement, sulfurization and desulfurization reactions, to generate an array of biomarker molecules (Tissot and Welte 1984). The functional groups are partly or entirely removed and the biomarkers can have different stereo- and structural isomers. Thus, the biomarkers indicate extent and relative speed of diagenetic chemical reactions such as redox state, pH, and availability of catalytic sites on mineral surfaces in the sediment during and after deposition. During catagenesis, biomarkers undergo structural changes that are used to measure the extent of heating of the source rocks or petroleum expelled from these rocks. At temperatures in the range of $\sim$150–200 °C, larger organic molecules are cracked to gas, thus biomarkers get severely diminished in concentration or are completely destroyed because of their instability in temperature range prior to greenschist metamorphism (Fig. 6.4).

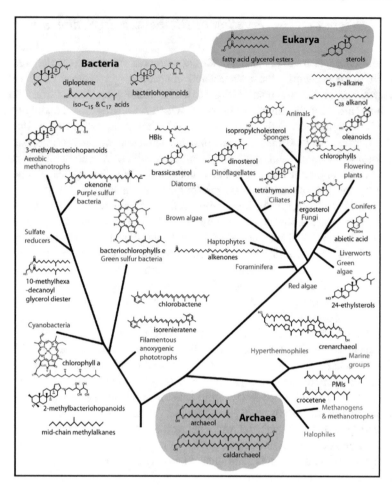

Fig. 6.3 Biomarker tree of life showing the three domains of life (adapted and modified from Briggs and Summons 2014)

6.3.2 Classification of Biomarkers

Nearly all biomarkers originate from two biosynthetic pathways that continue to exist since Proterozoic times on Earth (Hunt 1996). The first involves the enzyme controlled condensation of 2-carbon acetic acid structures (CH_3COOH) to form long carbon chains whose lengths are in multiple of two, e.g., C_{12}, C_{14}, and C_{16} etc. The second biosynthetic pathway involves polymerization of isoprene (methyl 1, 3 buta-diene), a 5-carbon building block (Fig. 6.4). It undergoes repetitive condensation via a compound isopentenyl pyrophosphate to form highly branched and cyclic iso-prenoids in multiples of five carbon atoms e.g. C_{10}, C_{15}, C_{20}, C_{25}, C_{30}, C_{35} and C_{40} called terpenoids, isoprenoids or isopentanoids (Hunt 1996) (Fig. 6.5).

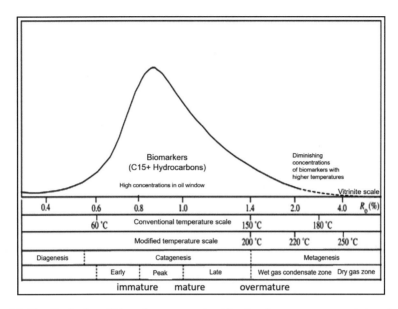

Fig. 6.4 The distribution of biomarker compounds with subsurface temperature variations (modified after Brocks and Summons 2003)

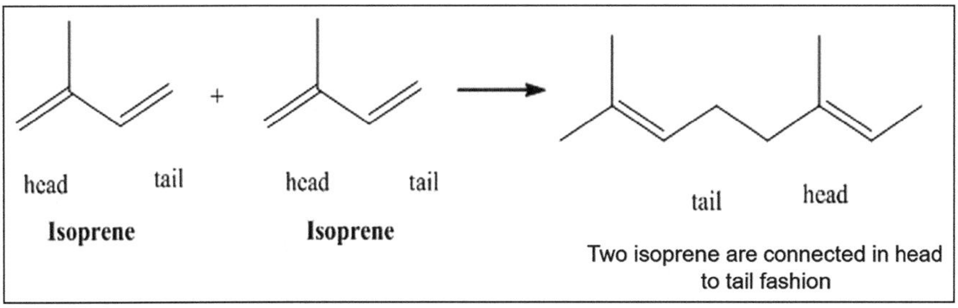

Fig. 6.5 Two isoprene units condensing to form a monoterpene (C_{10}) (modified after Hunt 1996)

Two monoterpanes (four isoprene units) link together to form a diterpane, whereas six isoprene units can be joined either to form a sterane or a triterpane, depending upon the linking pattern. The main compound classes are considered here with respect to origin and occurrence of specific molecules.

6.3.2.1 Saturate Biomarkers

The burial and diagenesis processes lead to the transformation of the original complex molecule involving loss of branches and multiple bonds and functional groups and make it primarily single bond (saturated) compound called alkanes. Alkanes are saturated hydrocarbons comprising of acyclic and cyclic biomarkers (Fig. 6.6)

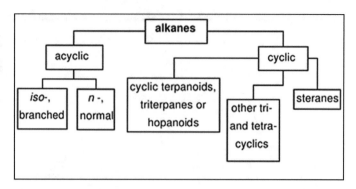

Fig. 6.6 Schematic of subgroups of alkane biomarkers

n-Alkanes, like hexadecane (C_{16}) are the most abundant hydrocarbons in all non-biodegraded oils and mature bitumens and its precursors can be found in all extant organisms such as bacteria and algae, algaenans of microalgae and waxes introduced by vascular plant debris (Brocks and Summons 2003). An interesting aspect of alkane biosynthesis is organism's ability to regulate the length of carbon chain and its unsaturation. Plants exclusively synthesize odd number carbon chains. Even number too are present but in very small proportion (Hunt 1996). Marine plants synthesize hydrocarbon compounds with odd carbon chain in the C_{15}–C_{21} liquid range, whereas the land plants synthesize odd carbon chain length in C_{25}–C_{35} solid range. *n*-Alkane profiles can be taxonomically and environmentally diagnostic (Mani 2019). The origin and maturity of organic-matter affects the predominance of odd and even carbon number. Carbon preference index (CPI) is the ratio obtained by dividing the sum of the odd carbon-numbered alkanes to the sum of the alkanes having even numbered carbon (Tissot and Welte 1984). The ratio is high (five-ten) in immature sediments and is about one in mature shales.

The chlorophyll molecule has phytol side chain which gives rise to varied ratios of acyclic isoprenoid markers, the pristanes and phytanes (Pr/Ph) for a particular sedimentary redox and is generally used as an indicator for different types of depositional environments (Fig. 6.7). High Pr/Ph ratios (<3) indicate contributions from terrestrial organic-matter of higher plants which have undergone some oxidation prior to preservation, whereas low Pr/Ph values (<2) indicate aquatic depositional environments including marine, fresh and brackish water (reducing conditions). The intermediate values (2–4) indicate fluvio-marine and coastal swamp environments and high values (up to 10) are related to peaty-swamp depositional environments (oxidizing conditions) (Tissot and Welte 1984). Acyclic alkanes of low molecular weight (C_{14}–C_{20}; monomethylalkanes) with one or more sites of branching are indicators of microbial mat communities, particularly those where cyanobacteria are predominant.

Terpenoids are cyclic isoprenoids and include C_{10} (mono-), C_{15}(sesqui-), C_{20}(di-), C_{30}(tri-), and C_{40} tetraterpenoids. Triterpanes contain 3–6 rings of which

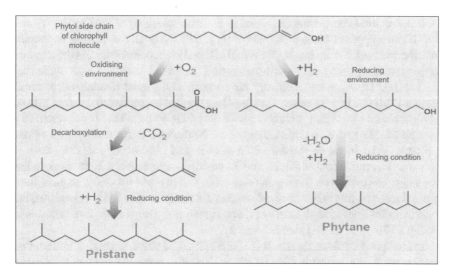

Fig. 6.7 Pristane and phytane markers from phytol chain of chlorophyll molecule as indicator of oxygenation conditions (modified after Peters et al. 2005)

Fig. 6.8 Structure of C_{35} homohopane showing carbon positions for important stereoisomeric configurations

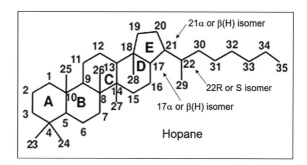

5-ring compound are most common. The E-ring usually contains five carbon atoms, as in hopane (Fig. 6.8), but compounds having six carbon atoms in the E-ring e.g. gammacerane are also known. Amongst the terpenoid group, the pentacyclic triterpenoids, including hopanes, are widely used for source rock characterization (Hunt 1996). The pentacyclic compounds are classified into two groups, hopanoids and non-hopanoids. Hopanes (the degraded and saturated hopanoid) are ubiquitous in sedimentary organic-matter and petroleum of all geological ages and are synthesized in nature from wide varieties of bacteria, cyanobacteria and other primitive organisms with prokaryotic cell. The characteristic base structure of a hopane comprises of four cyclohexane rings and one cyclopentane ring, and can have a side chain emerging from C_{30}. The C_{31}–C_{35} extended (side chain) hopanoids occur only in microorganisms. Hopanoids are also not present in methanogen or other archaebacterial (Hunt 1996).

The structural complexity (R, S and α, β steroisomerism) of the hopanes allows it to be an important biological marker (Fig. 6.8). Naturally occurring hopanoids have thermodynamically less stable $17\beta(H)$, $21\beta(H)$ stereochemistry. As the organic-matter undergoes diagenesis and maturation, saturated Hopanes with more stable $17\alpha(H)$, $21\beta(H)$ stereochemistry are formed. Thus, most mature source rocks contain $17\alpha(H)$, $21\beta(H)$ isomer. The C_{31}–C_{35} extended hopanes also occur in $17\beta(H)$ configuration in immature sediments with only 22R epimer. This is converted to a mixture of 22S and 22R during diagenesis. Norhopane is formed by loss of one carbon at 30 positions on the side chain and bis- and tris- norhopanes from the loss of two and three carbons at 22, 29 and 30 positions, respectively. Homohopane has one additional carbon on the hopane side chain and bis- and tris-homohopanes have two and three additional carbons. Moretanes are not present in living organisms, but form from hopanes at high levels of thermal maturity. They differ from hopanes in having a 17β (H), 21α (H) stereochemistry.

Ts is a specific trisnorhopane. It is $18\alpha(H)$ 22,29,30 trisnorhopane. Tm is another specific trisnorhopane with 17α (H) 22,29,30 configuration. Ts is more stable than Tm, and thus the ratio of Ts to Tm shows maturity of a source rock (Hunt 1996). The presence of alkyl substituent on the hopanoid skeleton, such as A-ring methyl groups, appears to be limited to explicit physiological taxae.g. 2 methyl hopanes indicate of cyanobacterial contribution. Hopanes which have benzene ring condensed to the E ring are called benzohopanoids and are associated with carbonate-evaporite deposi-tional environment of restricted circulation. The nonhopanoids are also pentacyclic triterpenoids, but do not contain the characteristic hopane structure. Some of these are gammacerane, friedelane, olenane and lupanes. They originate from higher plants (Hunt 1996; Tissot and Welte 1984).

Steranes contain four rings, of which the D-ring always contains five carbon atoms. The tetracyclic steroids and their sterane derivatives are not terpenoids. They do not follow the isoprene rule and involve oxidation and decarboxylation steps during their synthesis which destroys the original isoprenoid structure of precursor squalenemolecule (Hunt 1996). Cholesterol is a widespread and important sterol in plants and animals since Proterozoic. The eight asymmetric carbons give rise to 256 different stereoisomers of cholesterol, which after diagenesis and maturation in sediments indicate the changes that follow to form a lesser strained and more stable isomer (Fig. 6.9).

After organism dies, sterols are converted to stanols, sterenes and finally steranes by means by microbial activity and low-temperature diagenetic reactions. Steranes, as such do not exist in living organisms, but their concentration increases with deeper burial in sediments. Most widely used steranes for source rock characterization have C_{27}–C_{29} carbon atoms. Marine sterols extending from C_{26}–C_{30} and sterane deriva-tives down to C_{19} have been identified in sediments and oils. Cholesterol is C_{27} sterol. Cholestane can be C_{27}–C_{30} and is a sterane without the R group on its chain. C_{27}–C_{28}–C_{29} ternary diagram is a useful indicator of depositional environment. Nor-cholestane is a cholestane with one carbon missing. The diagenesis of sterols leads to rearranged sterenes called diasterenes, which gradually reduce to diasteranes. Diasterane to sterane ratios are used to distinguish oils generated through clastic

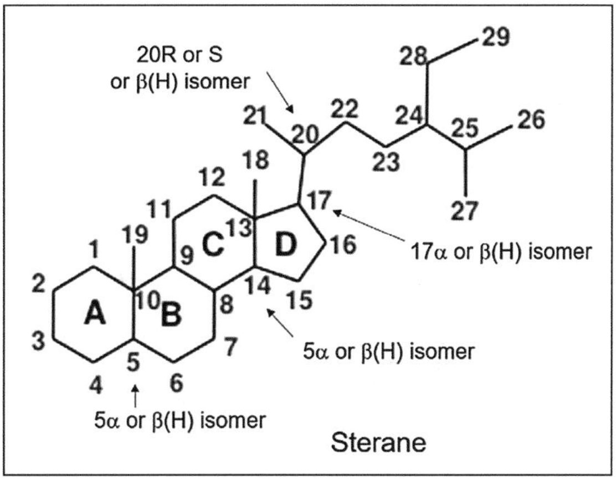

Fig. 6.9 Structure of C_{29} sterane molecule showing important stereoisomeric configurations (modified after Hunt 1996)

vs carbonate source rock and their stereoisomers ($\beta\beta/\beta\beta + \alpha\alpha$) are indicators of increasing thermal maturity.

6.3.2.2 Aromatic Biomarkers

The aromatic compounds found in source rocks are apparently derived from transformation through thermal degradation of steroids and terpenoids. Steroids give rise to substituted phenanthrenes and terpenoids produce alkylnaphthalenes (Mani et al. 2017). The distribution of aromatic hydrocarbons and their alkyl derivatives is strongly controlled by thermal maturation of organic-matter in the source rocks. They are resistant to biodegradation and respond to an increase in thermal stress with a predictable alkylation progression of a given parent compound or a shift in the isomer distribution of alkyl-aromatic homologues towards thermally more stable isomers (Horsfield and Schulz 2012). The α-substituted isomers of the polynuclear aromatic hydrocarbons are sterically hindered to a greater extent than the β-substituted isomers and are thermodynamically less stable. The ratios of α/β isomers being maturity dependent, their concentrations are applied in the calculation of temperature-sensitive maturity parameters such as methylphenanthrene indices (MPI), dibenzothiophene (DBT) ratiosand di, tri and tetra methylnaphthalene ratios (Table 6.1).

Some steranes form monoaromatic A rings and C rings during diagenesis. Both have been detected in immature sediments, but only the C-ring monoaromatics are prominent in mature source rocks (Fig. 6.10).

Aromatization of the B ring has been observed only after the steroids rearrange to an anthrasteroid (linear fused ring) structure (Hunt 1996). C-ring monoaromatic steroid occur in two groups, C_{27}–C_{29} and C_{20}–$C_{21,}$ the former containing longer side chain. The thermal maturation results in side-chain cleavage to give the smaller molecule. There are also C_{21} and C_{22} steroid natural products that can form short side chain monoaromatic steroids. The triaromatic steroids do not occur in recent

Table 6.1 Important aromatic marker indices for source rock maturity

Aromatic markers	Ratios
Methylnaphthalene ratio, MNR	2-MN/1-MN
Dimethylnaphthalene ratio-1, DNR-1	(2,6-DMN + 2,5-DMN)/1,5-DMN
Trimethylnaphthalene ratio-1, TMNr	1,3,7-TMN/(1,3,7-TMN + 1,2,5-TMN)
Tetramethylnaphthalene ratio, TeMNr	1,3,6,7-TeMN/(1,3,6,7-TeMN + 1,2,5,6-TeMN + 1,2,3,5-TeMN)
Pentamethylnaphthalene ratio-1, PMNr	1,2,4,6,7-PMN/(1,2,4,6,7-PMN + 1,2,3,4,5-PMN)
Methylphenanthrene index-1, MPI-1	1.5(2-MP + 3-MP)/(P + 1-MP + 9-MP)
Methylphenanthrene ratio, MPR	2-MP/1-MP
Methyldibenzothiophene ratio, MDR	4-MDBT/1-MDBT
Methyldibenzothiophene ratio, MDR	4-MDBT/(4 + 1)-MDBT

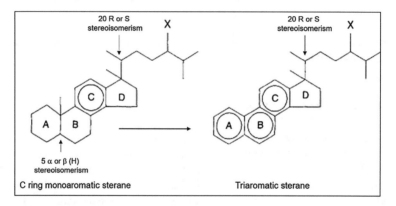

Fig. 6.10 Mono- and tri aromatic steranes indicating important stereoisomeric positions

sediments. They are formed from monoaromatic steroids with increasing depth of burial in response to increasing thermal maturation process.

The basis of biomarkers such as steranes and hopanes being source environment indicators stems from fundamental fact that eukaryotic cells utilize sterols in their cell walls, whereas prokaryotic cells utilize hopanoids. Thus, their ratios reflect input from eukaryotic (algae and higher plants) versus prokaryotic (bacteria) organisms to the source rock. Related source organic input of different thermal maturity will fall along a line on plots of sterane versus hopane concentration, whereas unrelated one with different eukaryotic versus prokaryotic input, may not follow it (Hunt 1996).

6.4 Biomarkers in Shale Source and Reservoir Rocks

Organic richness and thermal maturity are two important criteria for the characterization of prolific and potential shale rocks for gas generation prospect (Horsfield
and Schulz 2012; Jarvie et al. 2015). The natural gas is stored in shale gas reservoirs in fractures and pores, and absorbed on organic-matter. High total organic
carbon (TOC) content together with high thermal maturity of the organic-matter
increases the possibility for gas generation as well as absorption of the methane in
shales. Methane generation in shale source rocks is also a function of organic-matter
types. The H/C ratios of organic-matter from different sources vary, with lacustrine
and marine organic-matter having higher values compared to the terrestrial organic-
matter. This has implications on the gas generation potential of differently sourced
organic-matter (Table 6.2).

The successful shale source and reservoirs studies have defined few characteristics
typical of shale gas occurrence (Horsfield and Schulz 2012; Jarvie et al. 2015).
These include: (i) shale, siltstone or mudstone lithologies and mineralogy; (ii) TOC
greater than 2 wt%; (iii) thickness of shales greater than 30 m; (iv) thermal maturity
in the wet gas window (0.8–1.2% vitrinite reflectance equivalent) or the dry gas
window greater than 1.2% vitrinite reflectance equivalent; and (v) OM is not oxidised
(Jarvie et al. 2015). Biomarker studies help to define source (Table 6.3) and thermal
maturity (Table 6.4) of source rock and provide information about post-depositional
processes that may have affected it. It also provides important information related to
environmental hazards of shale gas production in an area.

The temperature sensitive biomarkers are influenced by the thermal stress developed during the burial and maturation of sedimentary organic-matter. These biomarkers indicate the thermal maturity of given organic-matter in a shale source rock.

Table 6.2 Oil and gas generation potential of organic-matter from varied sources

Depositional environment	Kerogen Type	Origin	Oil and gas potential
Aquatic	I	algal bodies	+ ↑
		structureless debris of algal bodies	
	II	structureless, planktonic material, primarily of marine origin	Oil
		skins of spores and pollen, cuticle of leaves, herbaceous plants	
Terrestrial	III	fibrous and woody plant fragments and structureless colloidal humic matter	Gas and some oil
	IV	oxidized, recycled woody debris	none −

Table 6.3 Source specific biomarkers and their proxy information

Biomarkers	C range	Proxy information
n-alkanes		
CPI > 5	C_9–C_{21}	Marine, lacustrine algal source, C_{15}, C_{17} and C_{19} dominant
	C_{25}–C_{37}	Terrestrial plant wax source, C_{27}, C_{29}, C_{31} dominant
CPI < 1	C_{12}–C_{24}	Bacterial source: oxic, anoxic, marine, lacustrine
	C_{20}–C_{32}	Saline or Anoxic environments: Carbonates and evaporites
Acyclic isoprenoids		
Head to tail		
Pristane	C_{19}	Chlorophyll, oxic, suboxic environments
Phytane	C_{20}	Chlorophyll, anoxic, saline
Head to head	C_{25}–C_{30}, C_{40}	Archaebacteria
Botrycoccane	C_{34}	Lacustrine, Brackish
Sesquiterpenoids		
Cadalene, eudesmane	C_{15}	Terrestrial plants
Diterpenoids		
Abietane, pimarane, kaurane, retene	C_{19}–C_{20}	Higher plant resins
Tricyclic terpanes	C_{19}–C_{45}	Degradation products of cell wall lipids of bacteria and alga
Tetracyclic terpanes	C_{24}–C_{27}	Degradation of pentacyclic triterpenoids
Hopanes	C_{27}–C_{40}	Bacteria
Norhopanes	C_{27}–C_{28}	Anoxic marine
2- and 3-methyl hopanes	C_{28}–C_{36}	Carbonate rocks
Benzohopanoids	C_{32}–C_{35}	Carbonate environments
Hexahydrobenzohopenoids	C_{32}–C_{35}	Anoxic, Carbonate-anhydrite depositions
Gammacerane	C_{30}	Hypersaline environment
Oleananes, lupanes	C_{30}	Products of Late Cretaceous and Tertiary flowering plants
Bicadinane	C_{30}	Resins from Gymnosperm tree
β-carotane	C_{40}	Arid, hypersaline environments
Steranes	C_{19}–C_{23}	Eukaryote organisms, plants and animals
	C_{26}–C_{30}	
24-n-propylsteranes	C_{30}	Restricted to marine sediments
4-methylsteranes	C_{28}–C_{30}	Marine and lacustrine dinoflagellates
Dinosteranes	C_{30}	Marine, Triassic or younger rocks

Table 6.4 Temperature sensitive biomarker ratios

Organic fractions	Biomarker Parameter Measured	Effect of Increasing Maturity	Interpretations
	C_{29} Steranes [20S/(20S + 20R)]	Increase	Ratios high in early to mid-oil window. Decreases at very high maturity levels
	C_{29} Steranes [$\alpha\beta\beta$/($\alpha\beta\beta$ + $\alpha\alpha\alpha$)	Increase	Ratios important in early to mid-oil window
Saturated	Moretane/hopane	Decrease	Useful in early oil window
Hydrocarbons	C_{31} Hopane [22S/(22S + 22R)	Increase	Useful in immature rocks to onset of early oil window
	Ts/(Ts + Tm)	Increase	Also influenced by source lithology
	Tricyclic terpanes/hopanes	Increase	Useful in late oil window; also increases at high levels of biodegradation
	Diasteranes/steranes	Increase	Useful in late oil window; also affected by source lithology (low in carbonates, high in shales); also increases at high levels of biodegradation
	Monoaromatic steroids: (C_{21} + C_{22})/[C_{21} + C_{22} + C_{27} + C_{28} + C_{29})	Increase	Useful in early to late oil window; resistant to biodegradation
Aromatic hydrocarbons	Triaromatic steroids: (C_{20} + C_{21})/[C_{20} + C_{21} + C_{26} + C_{27} + C_{28}]	Increase	Useful in early to late oil window; resistant biodegradation
	Triaromatic/(monoaromatic + triaromatic steroids)	Increase	High early to late oil window; resistant to biodegradation

6.5   Stable Isotope Variations in Sedimentary Organic-Matter

Stable isotopes of lighter elements like carbon, hydrogen, nitrogen and oxygen have been used to study diverse aspects of organic-matter burial, productivity, storage and cycling on Earth. Due to kinetic isotope effects (KIE), the transformation of inorganic carbon (CO_2) into living matter (C_6 carbohydrate) entails a noticeable bias in favour of lighter isotopes (^{12}C), with heavier species (^{13}C) retained in the inorganic reservoir (Schidlowski 1987). The isotopic discrimination leads to the preferential accumulation of ^{12}C in all forms of biogenic or reduced carbon as compared with the inorganic or oxidized carbon reservoir of surfacial environment, mainly in form of the atmospheric carbon dioxide and dissolved marine carbonate (Schidlowski 1987). When the biogenic materials and carbonates are incorporated in recent sediments, the KIE associated with the autotrophic carbon fixation (photosynthesis) is propagated from the surfacial exchange reservoir into the rock section of the carbon cycle (Schidlowski 1987). Over billions of years of Earth's history, this effect has ultimately brought a conspicuous isotope disproportionation (fractionation) of earth's primordial carbon into a light and heavy crustal carbon reservoir. The isotopic variations exhibited in the organic-rich shales provide information on the productivity, gross carbon storage and development of prolific source rocks in sedimentary basins over the geological time (Sharp 2017) (Fig. 6.11).

The hydrocarbon components generated from different types of organic-matter differ in isotopic compositions due to the characteristic parent source material. Depending upon the different enzymatic process used for carbon fixation in photosynthesis, there may even be different isotopic values for terrestrially derived organic-matter, in addition to the variability of marine and lacustrine organic signatures. Plants using C_3 Calvin cycle show less efficient carbon fixation and observe

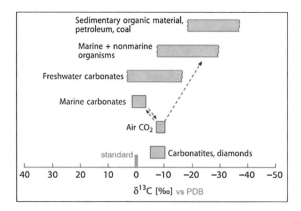

Fig. 6.11 Range of carbon isotope values in different reservoirs (modified after Sharp 2017)

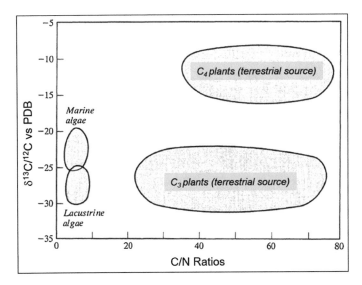

Fig. 6.12 Carbon and nitrogen isotope ranges for organic-matter sourced from different biological mass (adapted and modified after Sharp 2017)

large ^{13}C depletions compared to plants adopting C_4 cycle (Hoefs 2004; Sharp 2017) (Fig. 6.12).

Bulk and compound specific stable isotopes, specifically for carbon compounds are generally studied for identifying the origin of hydrocarbon gases from specific parent organic type, the maturity of source rocks, post-generation alteration, and gas-source relationships (Barbara and Karlis 2013). The bulk and compound specific stable can be studied in the whole rocks, extracted kerogen or organic-matter, direct gas and the gas desorbed or evolved from pyrolysis experiments.

6.5.1 Stable Isotopes Ratios of Natural Gas and Biomarkers

Depending upon the respective source of organic-matter, the originating gas systems have been classified into two distinct types, namely biogenic (or microbial) and thermogenic (Fig. 6.13). The biogenic (microbial) gases are formed in swampy areas, marine and immature sediments, and in the near-surface environments. These gases are generated at low temperature and shallow depths (diagenesis) and have heavier isotope depleted (^{13}C; $<-60‰$) and low to negligible concentrations of C_{2+} (ethane and other higher hydrocarbons). At greater subsurface depths and temperatures, thermal decomposition of organic-matter (catagenesis) yields natural gases which are often associated with petrogenic oil and gas. The thermogenic gases are characterized by an enrichment of ^{13}C ($>-60‰$) with high gas wetness ratio ($C_2–C_5/C_1–C_5 > 1$) (Schoell 1983). There can also be mixed gases from the two gas

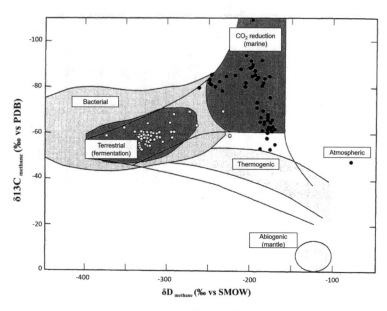

Fig. 6.13 Varied sources of methane based on $\delta^{13}C$ and δ^2H isotopes (adapted and modified after Sharp 2017)

types. Presence of subsurface thermogenic gas establishes the possibility or existence of source rock/organic rich horizons and provides information on the elements of the petroleum or gas system, which otherwise lacks when biogenically produced gases are encountered (Donna 2017; Mani 2019).

The North American shale gas are characterised by mainly methane with a minor amount of ethane and propane and minimal non-hydrocarbon gases. The stable isotopic composition of gases varies as a function of depth, thermal maturity, and gas wetness in unconventional shale gas accumulations and has distinguished features, defined as isotope reversals or rollovers (Barbara and Karlis 2013). The Carbon isotope reversal is the enrichment of methane $\delta^{13}C_1$ followed by other alkanes ($\delta^{13}C_1 > \delta^{13}C_2$ or $\delta^{13}C_2 > \delta^{13}C_3$), which is opposite of that observed in conventional petroleum source rocks. It has been a common occurrence at high level thermal maturity areas and in some cases; these features identify the most productive intervals within shale gas systems. Ethane and propane isotope rollovers indicate in situ cracking at high maturities and demonstrate overpressured zones (Barbara and Karlis 2013; Wood and Hazra 2017).

The organic-matter extracts, referred to as extracted organic-matter (EOM) are often studied for its stable bulk and compound specific isotope analysis (CSIA) of carbon and hydrogen. The prime advantage of CSIA over bulk is that interpretation of bulk isotope values can be confounded by variations and overlapping of isotopic ratios, which becomes better resolved and constrained when the carbon isotope ratios of the individual alkanes or biomarker compounds are determined. CSIA increases the molecular specificity, through which the primary versus secondary and allochthonous versus autochthonous materials in the shale organic-matter can be distinguished. The marine and continental (terrestrial and freshwater) plants show different δ^{13}C signatures due to the difference in the isotopic composition of carbon sources. The depositional environment of marine versus non-marine shales has a direct influence on the type and amount of organic-matter that they contain and on their gas generation potential.

6.6 Analytical Approaches

The analytical approach used to study the properties of organic-matter involvesits extraction and separation into different fractions of saturates, aromatic and polar compounds. This is followed by its qualitative and quantitative analysis on Gas chromatograph–mass spectrometer (GC–MS) and stable isotope ratio determination over isotope ratio mass spectrometry (IRMS).

6.6.1 Gas Chromatography–Mass Spectrometry (GC–MS)

Gas chromatography–mass spectrometry (GC–MS) is a separation technique of (GC) hyphenated with mass spectrometry (MS). The relative gas chromatographic retention times and elution patterns of components of a mixture are used in combination with the characteristic mass spectral fragmentation patterns of a compound (Sneddon et al. 2007). A GC–MS system has functions: (1) it separates the individual compounds in a mixture through the column of a gas chromatograph; (2) transfers the separated components to the ionizing chamber of mass spectrometer; (3) performs ionization; (4) mass analysis; (5) detection of the ions using an electron multiplier; and 6) data acquisition, processing, and display using computer system (Fig. 6.14).

   After elution from the GC column, organic compounds enter the ionization chamber of MS. where they are bombarded by a stream of electrons leading to the fragmentation into ions. The mass to charge ratio (m/z) is characteristic of the molecule and represents the molecular weight of the fragment. There can be magnetic sector or quadrupole type mass analysers to detect the respective ions. A mass spectrum results, which is a graph of the signal intensity (relative abundance) versus the m/z ratios (essentially the molecular weight) (AOGS 2019). The homologous series of saturated and aromatic fractions are identified by characteristic masses of the fragmented ion and their retention times in the chromatographic column e.g. alkanes by $m/z = 57$; steranes by $m/z = 271$; hopanes by $m/z = 191$ etc. The individual compounds or their isomers in a series are identified by the characteristic fragmentation pattern of the molecular ions, its retention time, and use of respective biomarker standards and/or published mass spectra (Fig. 6.15).

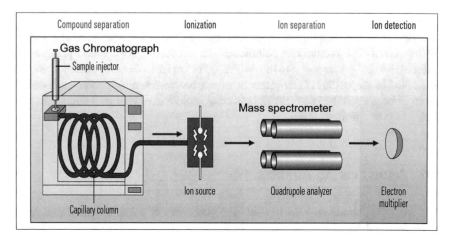

Fig. 6.14 Components of a Gas Chromatograph-Mass Spectrometer (adapted and modified from Mani 2019)

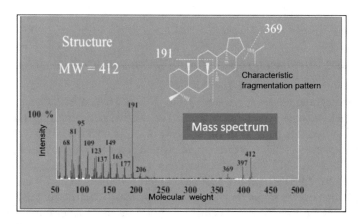

Fig. 6.15 A mass spectrum of αβ hopane showing base peak at m/z = 191 and molecular ion peak at m/z = 412

6.6.2 Isotope Ratio Mass Spectrometry (IRMS)

Isotope ratio mass spectrometry (IRMS) is used to determine the ratio of stable isotopes of low molecular weight elements such as carbon ($^{13}C/^{12}C$) and oxygen ($^{18}O/^{16}O$). Such studies are performed using coupled techniques of a gas chromatograph and the mass spectrometer, along with the introduction of a sample combustion interface into the gas chromatograph-IRMS (Platzner, 1997). In GC–C–IRMS, the separated products of the sample mixture, carried in the stream of helium, are eluted at the output of the gas chromatograph and passed through an oxidation/reduction reactor and then are introduced into the ion source of mass spectrometer for the ratio

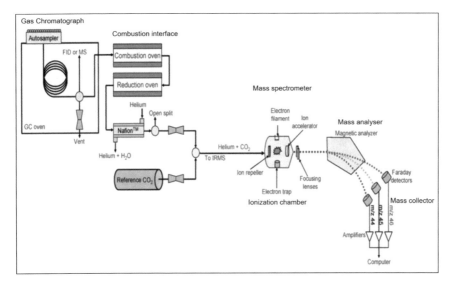

Fig. 6.16 A schematic of Gas Chromatograph-Combustion-Isotope Ratio Mass Spectrometer (modified from Mani 2019)

determination (Fig. 6.16). An open split-coupling device ensures that only a part of the sample/reference gas containing carrier gas is fed into the ion source, thus reducing the volume constraints and sample size.

Continuous flow-IRMS is a standard term used for the IRMS instruments that are coupled online to preparation lines or instruments. This includes the: (1) gas chromatography–combustion–IRMS (GC–C–IRMS), used for the compound specific isotope ratio determination of individual hydrocarbon components, (2) gas bench–IRMS (GB–IRMS), used for the C and O isotope ratio determinations on carbonates, and (3) elemental analyzer–IRMS (EA–IRMS), for the determination of bulk isotopic compositions of the coexisting organic-matter.

References

AOGS (2019) Australian quantitative analysis of Petroleum Biomarkers using the AGSOSTD oil. http://www.ga.gov.au

Barbara T, Karlis M (2013) Isotope reversals and universal stages and trends of gas maturation in sealed, self-contained petroleum systems. Chem Geol 339(15):194–204

Briggs DEG, Summons RE (2014) Ancient biomolecules: their origins, fossilization, and role in revealing the history of life. BioEssays 36:482–490

Brocks JJ, Summons RE (2003) Sedimentary hydrocarbons, biomarkers for early life. In: Holland HD (ed) Treatise in geochemistry, vol 8, 53p. Elsevier

Donna CW (2017) Aromatic compounds as maturity indicators comparison with pyrolysis maturity proxies and ro (measured and calculated) using the New Albany Shale as an example search and discovery article #42143

Hallmann C, Kelly AE, Gupta SN, Summons RE (2011) Reconstructing deep-time biology with molecular fossils. Chapter in 'Quantifying the evolution of early life', pp 355–401. Springer, Dordrecht

Hoefs J (2004) Stable isotope geochemistry. Springer, Berlin, p 244

Horsfield B, Schulz HM (2012) Shale gas exploration and exploitation. Mar Pet Geol 31(1):1–2

Hunt JM (1996) Petroleum geology and geochemistry. W. H. Freeman and Company, San Francisco, p 617

Jarvie DM, Jarvie BM, Weldon WD, Maende A (2015) Geochemical assessment of in situ petroleum in unconventional resource systems. In: Unconventional resources technology conference. Society of Petroleum Engineers, San Antonio

Mani D (2019) Characterising the source rocks in petroleum systems using organic and stable isotope geochemistry: an overview. J Indian Geophys Union 23(1):10–27

Mani D, Kalpana MS, Patil DJ, Dayal AM (2017) Organic-matter in gas shales: origin, evolution and characterization. In: Shale gas: exploration, environmental and economic impacts, vol 3. Elsevier, pp 25–52

Peters KE, Walter CC, Moldowan JM (2005) The biomarker guide, vol 1: biomarkers and isotopes in the environment and human history. Cambridge University Press

Platzner IT (1997) Modern isotope ratio mass spectrometry. Wiley, Chichester

Schidlowski M (1987) Application of stable carbon isotopes to early biochemical evolution on earth. Annu Rev Earth Planet Sci 15:47–72

Schoell M (1983) Genetic characterization of natural gases. AAPG Bull. 67:2225–2238

Sharp Z (2017) Principles of stable isotope geochemistry, 2nd edn. Retrieved from https://digitalrepository.unm.edu/unm_oer/1/

Sneddon J, Masuram S, Richert JC (2007) Gas chromatography-mass spectrometry-basic principles, instrumentation and selected applications for detection of organic compounds. Anal Lett 40(6):1003–1012

Tissot BP, Welte DH (1984) Petroleum formation and occurrence. A new approach to oil and gas, 2nd edn. Springer

Wood DA, Hazra B (2017) Characterization of organic-rich shales for petroleum exploration & exploitation: a review—part 2: geochemistry, thermal maturity, isotopes and biomarkers. J Earth Sci 28(5):758–778

Chapter 7
Organic and Inorganic Porosity, and Controls of Hydrocarbon Storage in Shales

A significant portion of the gas/oil present in shale-reservoirs exists in adsorbed form within the porous structures present within shales (Curtis 2002; Zhang et al. 2012). Proper characterization of shale-porosity is thus essential for estimating and extracting the gas/oil present within them (Wood and Hazra 2017). In recent years, different techniques have been applied for understanding and evaluating the pore-spectrum of shales, which has helped in understanding their structure, geometry, distribution, evolution, and controlling parameters (Loucks et al. 2009; Curtis et al. 2012; Milliken et al. 2013; Löhr et al. 2015). Among the different techniques, low pressure gas adsorption (LPGA) and electron microbeam imaging or scanning electron microscopy (SEM) are the most widely used techniques (Wang and Reed 2009; Desbois et al. 2009; Cardott et al. 2015).

Curtis et al. (2012), using focused-ion-beam (FIB) milling and scanning electron microscopy (SEM) discussed the formation and evolution of porosity in Woodford Shale samples ranging in maturity from ~0.5% to >6% Ro. Organic porosity was found to be more abundant in shale samples with vitrinite reflectance values greater than ~0.90%. Loucks et al. (2009) also observed the presence of organic-matter hosted pores for Barnett Shale samples with >0.8% vitrinite reflectance. These and several other studies (Bernard et al. 2012a, b; Jennings and Antia 2013; Chen and Jiang 2016) have documented much lower porosity within thermally immature shales, while the creation of secondary porosity occurs within organic-matter as thermal maturity levels increase. However, the creation of porosity with increasing thermal maturity levels may not be a linear process. For example, Jennings and Antia (2013) also observed minimal porosity in shale samples with >1.7% Ro. Curtis et al. (2012) in their work observed the absence of porosity within a sample with ~2% Ro. However, at higher thermal maturity levels they did observe the presence of pores within the organic-matter present within the shales. On the other hand, Löhr et al. (2015) reported significant porosity within solvent-extracted thermally immature Devonian Woodford shales. They opined that bitumen held within pores, obscures the pores and leads them to be unidentified during electron-imaging techniques.

Pore-development in shales is also likely to be dependent on the type of maceral present within them (Curtis et al. 2012). The reactive, pyrolyzable and convertible

© Springer Nature Switzerland AG 2019
B. Hazra et al., *Evaluation of Shale Source Rocks and Reservoirs*,
Petroleum Engineering, https://doi.org/10.1007/978-3-030-13042-8_7

organic-carbon fraction (i.e., kerogen types I, II, and III) generate petroleum and leave a solid residue of dead or inert organic-carbon with increasing thermal maturity (Jarvie et al. 2007). Organic-matter-hosted nano-porosity is formed as part of the reactions converting kerogen into petroleum, resulting in the formation of liquids and gases that coalesce into bubbles (Loucks et al. 2009). On the other hand, inertinite macerals (e.g., fusinite and semifusinite) may have primary or inherent macroporosity (e.g., plant cell lumens). Liu et al. (2017) observed the formation of organic-pores within the alginite maceral as petroleum was produced from it during thermal maturation.

The role of inorganic-porosity in shales in providing petroleum storage and migration pathways is somewhat controversial (Curtis et al. 2012). Several researchers have documented that porosity hosted by the inorganic minerals and matrix in shales plays an important role in both petroleum storage and the flow (recovery/secondary migration) of petroleum from them (Curtis et al. 2010; Milner et al. 2010). Camp and Wawak (2013) examined the different perspectives of organic-hosted porosity versus inorganic-hosted porosity in shale reservoirs based on SEM analysis to highlight their distinctive morphologies and relative significance.

7.1 Low-Pressure Gas Adsorption

Evaluation of the complex pore size distribution (PSD) and pore sizes of shale reservoirs using subcritical gas-adsorption techniques, have gained much popularity in recent years (Ross and Bustin 2009; Chalmers et al. 2012; Mastalerz et al. 2012, 2013; Kuila and Prasad 2013). Nitrogen is the most commonly used gas for low-pressure and low-temperature gas adsorption experiments of shales. Using cryogenic liquid nitrogen at a constant temperature of -197.3 °C and relative pressure range of near zero to 1.0, a range of pore sizes (1.7–<200 nm) existent within shales can be studied (Kuila and Prasad 2013).The International Union of Pure and Applied Chemistry (IUPAC), based on their sizes classifies pores as—micropores (pore width ≤ 2 nm), mesopores (pore width between 2 and 50 nm), and macropores (pore widths > 50 nm) (Sing et al. 1985). Thus the limitation of using nitrogen gas is its inability to access all micropores and macropores having pore widths greater than 200 nm. For micropore evaluation, commonly CO_2 is used at low pressures and at 0 °C temperatures, and has been employed by several researchers (Ross and Bustin 2009; Mastalerz et al. 2012).

Low pressure gas adsorption (LPGA) experiments conducted using nitrogen as adsorbate at constant temperature, involves systematic increase in relative pressure levels (P/P0) [where P represents gas-vapor-pressure, while P0 represents the saturation-pressure of the adsorbent]. This gives the volume of gas-adsorbed at different P/P0 intervals, and thus helps in constructing the adsorption isotherms (AI). The AI generated offers a qualitative evaluation of the pore-structure of the sample being studied (Brunauer et al. 1940). Shape of AI and the hysteresis pattern helps in predicting the pore-types present within the rock. In general six-types of AI and

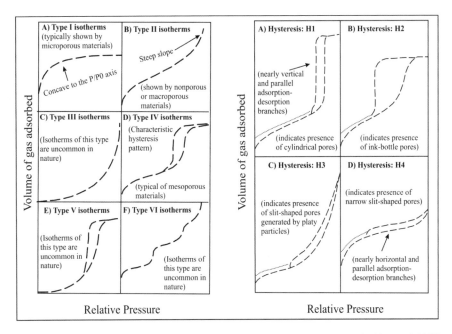

Fig. 7.1 Different shapes of isotherms and hysteresis patterns, as given by IUPAC (Sing et al. 1985)

4-types of hysteresis pattern are recognized by IUPAC (Fig. 7.1; Sing et al. 1985). Monson (2012) provides a review of hysteresis in the context of pore size analysis.

7.2 Specific Surface Area (SSA) and Pore Volumes of Shales

Multi-point Brunauer–Emmett–Teller (BET) (Brunauer et al. 1940) and Langmuir AI analysis (Langmuir 1918) are typically used for pore surface area measurements, making the assumptions of multi-layer and mono-layer adsorption, respectively. The mathematical basis for calculating pore surface area using these methods is described in detail by Gregg and Sing (1982). The well-established method for determining the specific surface area (SSA) of porous solids (BET) method. The Langmuir isotherm assumes mono-layer adsorption across an energetically uniform pore surface, where N_m is the number of adsorbate molecules required to cover the solid surface of the pore wall with a single mono-layer. These assumptions represent unrealistic conditions that typically do not provide a true measure of pore surface area. On the other hand, the BET AI assumes multi-layer adsorption in which the first layer follows the Langmuir AI assumptions, but subsequent adsorbed layers involve condensation of gas onto the liquid forming the underlying layer (i.e., not the solid pore wall) and heat of adsorption of the first adsorbed layer is greater than that of the subsequent overlying adsorbed layers. BET theory uses a transform plot to determine the statistical 'monolayer capacity', which is the amount (moles) of nitrogen required to cover the total surface area of the pore space (i.e., the surface area associated with meso-porosity and macroporosity in the sample) with a complete monolayer of nitrogen molecules.

The BET isotherm is expressed by Eq. (7.1), which reverts to the Langmuir equation when the number of layers is reduced to 1:

$$\theta = \frac{N}{N_m} = \frac{C(P/P_0)}{(1 - P/P_0)\left[1 - \frac{P}{P_0} + C(P/P_0)\right]} \tag{7.1}$$

where: θ represents the fraction of the pore surface covered or occupied with adsorbate. However, the BET isotherm is more usefully expressed for specific surface area measurements as Eq. (7.2) Trunsche (2007):

$$\frac{1}{W((P_0/P) - 1)} = \frac{1}{W_m C} + \frac{C - 1}{W_m C}\left(\frac{P}{P_0}\right) \tag{7.2}$$

where

W = weight of gas adsorbed
P/P_0 = relative pressure which typically varies between $0.05 < P/P_0 < 0.3$ for complete mono-layer coverage
W_m = weight of adsorbate as monolayer
C = BET constant.

This equation defines a linear plot (Fig. 7.2) of $1/[W(P/P_0) - 1]$ versus P/P_0 with the slope (s) of the line defined by Eq. (7.3):

$$S = \frac{C - 1}{W_m C} \tag{7.3}$$

Fig. 7.2 The linear BET relationship (modified after Trunsche 2007)

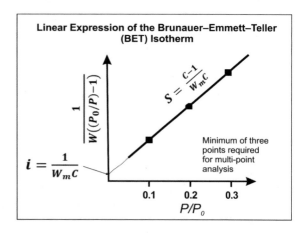

The intercept (i) is defined by Eq. (7.4)

$$i = \frac{1}{W_m C} \tag{7.4}$$

The weight of the mono-layer (W_m) is then derived by Eq. (7.5)

$$W_m = \frac{1}{(s + i)} \tag{7.5}$$

The BET constant (C) is derived as Eq. (7.6). For the common adsorbents the C constant typically falls in the range 50–300.

$$C = (s/i) + 1 \tag{7.6}$$

Although linear relationships are typically derived, BET theory is an approximation that assumes homogeneous surfaces for the pore walls and homogeneous adsorbate-adsorbate layer interactions, whereas heterogeneities can lead to non-linearity in the BET plots at $P/P_0 < 0.05$.

Based upon the assumed linear relationship the mono-layer capacity is then converted into the total surface area (S_t) using Eq. (7.7):

$$S_t = \frac{W_m N A_{cs}}{M} \tag{7.7}$$

where

N is Avogadro's number (6.023×10^{23})

M is the molecular weight of the adsorbate (e.g. 28.0134 for N_2)

A_{cs} = Adsorbate cross sectional area (16.2 Å^2 at $-197.3\,^\circ\text{C}$ for Nitrogen).

The specific surface area (SSA) is then calculated by dividing the total surface area by the sample weight as expressed by Eq. (7.8):

$$SSA = \frac{S_t}{W} \tag{7.8}$$

The single point BET calculation determines SSA from a single point on the isotherm, typically assuming an intercept of zero for high values of C. The more reliable multi-point BET SSA calculation involves determining at least three points along the isotherm (Fig. 7.2).

An example calculation of SSA is provided in Table 7.1. See also calculations by Leddy (2012).

The Barrett–Joyner–Halenda (BJH) method is the most widely used to determine pore volumes and pore size distributions for shale samples. This method exploits the nitrogen adsorption and desorption isotherms to identify a dominant pore-size range in a sample (Barrett et al. 1951). The BJH method determines pore size distributions

Table 7.1 Example multi-point BET SSA calculation for a Lower Permian Barakar Formation shale from Jharia basin, India using Eqs. (7.2)–(7.8)

Relative pressure (P/P₀)	Volume @STP (cc/g)	1/[W((P₀/P) − 1)]
0.066996	1.3218	0.054326
0.083589	1.377	0.066242
0.108404	1.448	0.083969
0.132992	1.5122	0.101434
0.163123	1.5865	0.122863
0.182321	1.631	0.136707
0.206959	1.6847	0.154902
0.235316	1.7434	0.176514
0.256323	1.7893	0.192624
0.280997	1.8439	0.211952
0.305639	1.8973	0.232001
Slope: 0.739948 ± 0.005450		
Y-intercept: 0.003352 ± 0.001086		
Correlation coefficient, r = 0.999756		
C constant: 221.76		
Surface area: 5.86 m²/g		

of a mesoporous samples taking into account capillary forces by applying the Kelvin equation, iteratively.

The Kelvin equation can be expressed in a generalized form in Eq. (7.9) (Pirngruber 2016):

$$\ln \frac{P_{cap}}{P_{sat}} = -\frac{V_m \gamma}{RT} \cdot \frac{dA}{dV} \qquad (7.9)$$

where

P_{cap} is the capillary pressure of the adsorbate in a pore
P_{sat} is the saturation pressure of the adsorbate in a pore
V_m is the molar volume
dV/dA is the change in volume per change in interface area
γ is the surface tension of liquid nitrogen
R is the universal gas constant
T is absolute temperature of the adsorbate in degrees Kelvin.

The Kelvin equation provides a correlation between pore diameter and the pore condensation pressure making the assumptions: (1) that pores are of a general cylindrical shape; and, (2) that no fluid-pore wall interactions occur.

The dV/dA metric is dependent upon the pore geometry, such that:

- for a spherical pore dV/dA = $r/2$ (where r is the radius of the pore

- for a cylindrical pore dV/dA = r

for a slit-shape or elongated pore dV/dA = $2r$ or d (where d is the diameter of the narrower dimension).

By taking the pore geometry into account, the Kelvin equation can also be expressed in a generalized form in relation to the curvature of the pore as Eq. (7.10):

$$\ln \frac{P_{cap}}{P_{sat}} = -\frac{2}{r} \frac{V_m \gamma}{RT} \tag{7.10}$$

The Kelvin equation expressed in the form of Eq. (7.10) is useful for modelling the capillary forces acting within a pore during the condensation and evaporation of adsorbate within a pore during low pressure gas adsorption tests.

This is particularly important when considering partially filled pore spaces in which a liquid film (condensed adsorbate) exists adsorbed to the pore walls. In such a situation, the capillary forces in the pore act to lower the vapor pressure of the condensed liquid forming the liquid film. This results in the pressure at the concave side of an interface between the liquid film and the gaseous adsorbate filling the centre of the pore is higher than at the convex side of that interface. That differential pressure counteracts surface tension (γ) thereby acting to progressively collapse the interface of the liquid film towards the pore wall as the pressure drops (Fig. 7.3).

The net effect of such capillary action is that the pressure in the liquid film is lower than the gas pressure in the pore causing capillary condensation to occur at a lower pressure than condensation. This means that during desorption as the pressure declines in small steps (e.g., from step $n - 1$ to step n pressure declines from $Pn - 1$ to Pn) capillary evaporation occurs within a pore in a manner that is related to the pore's geometry, as described by Eq. (7.11).

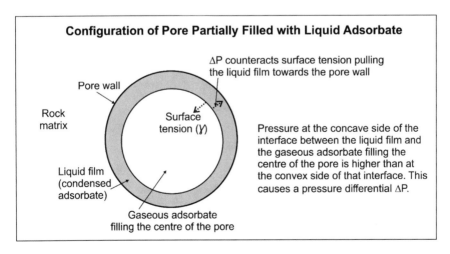

Fig. 7.3 Diagrammatic representation of capillary forces acting on the liquid film at the pore walls within a cylindrical pore

$$\Delta V = \frac{(r_k + \Delta t)^2}{r_p^2} \cdot V_p \tag{7.11}$$

where

V_p is the actual pore volume,
ΔV is the volume of adsorbate desorbed in the pressure step $Pn - 1$ to Pn. The reductions in the thickness of the liquid film in pore sizes that are already partially emptied contribute to the value of ΔV in each pressure reduction step during desorption.
r_p is the actual pore radius,
r_k is the radius of the portion of the pore filled with the adsorbate gas phase (radius of the capillary),
t is the thickness of the adsorbate liquid film covering the pore walls,
Δt is the change in the thickness of the liquid film during the pressure step $Pn - 1$ to Pn.

The evolving relationship between r_p, r_k and Δt through the desorption pressure steps is illustrated in Fig. 7.4 which explains the simple relationship between r_p, r_k and t expressed as Eq. (7.12)

$$r_p = r_k + t \tag{7.12}$$

At varying pressure differences along the desorption isotherm this capillary effect can be exploited to estimate the pore size distribution of a porous sample (Fig. 7.5).

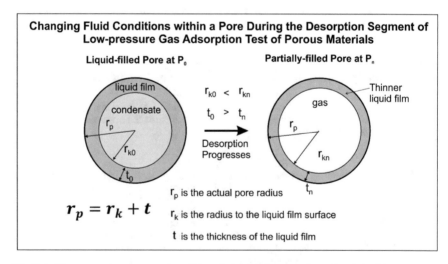

Fig. 7.4 Diagrammatic representation of the relationship between the radius (r_k) of the pore component filled with adsorbate gas and the actual pore radius (r_p)

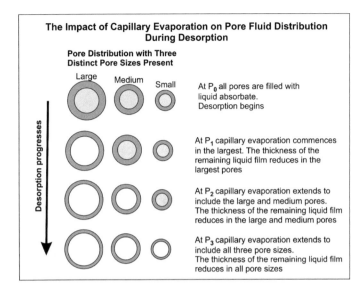

Fig. 7.5 Diagrammatic representation of the pore-size distribution model considered by the BJH method

For each desorption step, if the actual pore radius (r_p) is known, then the average diameter of the pore, which undergoes capillary evaporation (r_k) can be calculated by combining Eqs. (7.10) and (7.12) to form Eq. (7.13)

$$\log(P/P_0) = \frac{-2\gamma V_m}{8.316 * 10^7 * 2.303 T r_k} = \frac{-4.14}{r_k} \quad (7.13)$$

where

γ is the surface tension of liquid nitrogen,
V_m is the molar volume of nitrogen
r_k is the radius of the capillary (in cm units that are usually converted into nm or Å)
T is absolute temperature of the adsorbate in degrees Kelvin
8.316×10^7 is the universal gas constant (R) in ergs per degree.

The BJH method, for progressively calculating pore volume (V_{pn}) through each of n steps of desorption adjusts the capillary evaporation equation with a term that adjusts it for the progressively changing thickness of the liquid film for each pressure increment change. This is expressed as Eq. (7.14):

$$V_{pn} = \frac{r_{pn}^2}{(r_{kn} + \Delta t_n)^2} \cdot \Delta V_n - \frac{r_{pn}^2}{(r_{kn} + \Delta t_n)^2} \cdot \Delta t_n \cdot \sum_{j=1}^{j=n-1} \frac{r_{pj} - t_j}{r_{pj}} \cdot A_{pj} \quad (7.14)$$

where

r_k for each pressure step is derived from Eq. (7.13),
Δt_n is the change in layer thickness in each of n desorption steps,
ΔV_n is the volume desorbed in each of n desorption steps,
A_p is the surface area of each pore.

A_p is a constant for pores of a specified size and can be calculated for pressure increment of the desorption process by the geometric relationship expressed as Eq. (7.15)

$$A_p = 2\frac{V_p}{r_p} \tag{7.15}$$

This means that A_p can be summed for each successive step in the desorption process and thereby be successively computed as $\sum A_p$ (Barrett et al. 1951). The useful plot in characterizing the pore volume distribution of a shale that can be derived from Eq. (7.14) is pore volume versus pore radius.

The BJH method was developed with the desorption branch of the isotherm in mind. However, its method can also be usefully applied to the adsorption branch of the isotherm. Indeed, in some cases the desorption branch of the isotherm is characterized by artificial peaks or steps at certain pore diameters due to surface tension meniscus radius relationships. These adversely impact the reliability of the pore volume distribution calculations. In such cases it is more meaningful to use the smoother adsorption branch of the isotherm for pore volume distribution calculations.

The BJH method is an approximate method that is prone to underestimate micropore sizes as fluid-pore wall interactions are ignored and the Kelvin equation ceases to be realistic for small pores with high degrees of curvature. Alternative adjustments to the capillary evaporation equation are frequently applied. These include the modifications that take into account the influence of the meniscus radius of the adsorbed film on adsorbate surface tension, e.g. by Dubinin and Radushkevich (1947), elaborated by Stoeckli and Houriet (1976), and the statistical layer thickness of adsorbate, e.g. the Harkins-Jura (HJ) equation (Harkins and Jura 1944) or the Frenkel-Halsey-Hill (FHH) equation (Halsey 1948; Hill 1952; Gregg and Sing 1982).

Carbon dioxide gas has the ability to penetrate some of the smaller nanopores in rock formations at 0 °C and ambient experimental temperatures (Clarkson and Bustin 1999; Ross and Bustin 2009). Consequently, CO_2 adsorption isotherms can quantify of micropore volumes and surface areas of tight-rock formations with significant nanoporosity. Organic-rich shales act as a natural molecular sieve for carbon dioxide gas, allowing it to penetrate pores that other pure and natural gases are unable to reach. In particular, carbon dioxide is able to reside in an adsorbed state within nano-scale pores within kerogen (Kang et al. 2011).

Dubinin and Radushkevich (1947) proposed an equation to estimate the low and medium pressure parts of the adsorption isotherm. It involved a term defined as the differential molar work of adsorption (α) expressed as Eq. (7.16):

$$\alpha = RT \ln\left(\frac{P_0}{P}\right) \tag{7.16}$$

Their concept involved filling of micropores with the adsorbate rather than building up the liquid film on the pore walls layer-by-layer (the BJH concept). The fraction of micropore filling (θ) is a function of α.

The pore capacities for the carbon dioxide adsorption isotherm are often calculated by applying the Dubinin-Radushkevich (DR) equation expressed as Eqs. (7.17) and (7.18):

$$\frac{W}{W_0} = \exp\left[-B\left(\frac{T}{\beta}\right)^2 \cdot \log_{10}^2\left(\frac{P_0}{P}\right)\right] \tag{7.17}$$

$$B = 2.303\frac{R^2}{k} \tag{7.18}$$

where

W is the volume of adsorbed (cc/g) gas at equilibrium pressure,
W_0 is the total micropore volume (cc/g),
R is the universal gas constant,
T is absolute temperature in degrees Kelvin,
B is the structural constant of the adsorbent influenced by the pore structure of the adsorbent,
β is the affinity coefficient or similarity coefficient (a scaling factor), which is usually set at 0.46 at 273.15 K for CO_2, and,
k is a coefficient that expresses the breadth of the Gaussian distribution of the cumulative micropore volume divided by the normalized work of adsorption (α/β).

Equation (7.17) can be transformed into Eqs. (7.19) and (7.20):

$$\log_{10} W = \log_{10} W_0 - D \log_{10}^2\left(\frac{P_0}{P}\right) \tag{7.19}$$

$$D = B\left(\frac{T}{\beta}\right)^2 \tag{7.20}$$

According to Eq. (7.19), the DR plot of $\log_{10} W$ versus $\log_{10}(P_0/P)^2$ should be a straight line with its intercept representing the value W_0. The slope of that line represents the value B/β.

The carbon dioxide volumes derived from the Langmuir adsorption isotherm provide estimates of total gas capacity and have been shown to correlate linearly with DR micropore volumes, obtained from low-pressure carbon dioxide adsorption (Clarkson and Bustin 1999).

An alternative to the DR equation was proposed by Dubinin and Astakhov (1971), the DA equation, which modifies the DR equation by replacing the exponent of 2 with a small integer value n to provide an improved data fit. The exponent n is often assigned a value of 3 when applying the DA equation (Stoeckli et al. 1989). The DA equation can also be calculated to determine micropore surface area and limiting micropore volume. The DA isotherm in its general form is expressed by Eqs. (7.21) and (7.22):

$$W = W_0 \exp\left[-\left(\frac{A}{\beta E_0}\right)^n\right]$$
(7.21)

$$A = RT \cdot \ln\left(\frac{P_0}{P}\right)$$
(7.22)

where

E_0 is a characteristic of free energy of adsorption that is equivalent to α when $\theta = 1/e = 0.368$. The DR equation (Eq. 7.19) is a special case of the DA equation when $n = 2$. The coefficient B of the DR equation is related to E_0 by Eq. (7.23):

$$B = \left(2.303 \cdot \left(\frac{R}{E_0}\right)\right)^2$$
(7.23)

As is the case for the BET method, the DR and DA methods to analyze pore structure, were originally designed for homogeneous materials not the typically complex pore size distributions encountered in organic-rich shales (Li et al. 2016a, b). The fact that organic-rich shales are far from homogeneous, with porosity constituted by nanopores, micropores, mesopores and macropores, means that these models can only approximate the pore characteristics of specific shale samples. Li et al. (2016a, b) demonstrated for shale samples from the south east Chongqing area (China) that the method of Stoeckli et al. (1989) applying an alternative generalization to the DA equation to the CO_2 adsorption isotherm can provide simple pore size distribution (PSD) analysis of micropores in shales. Their overall isotherm sums the contributions of specific pore-size groups with each group characterized by its own W_0 and B_j values and obeying the DA equation. Equations (7.24) and (7.25) represents the specific DA representation applied to each group of pore sizes using an n exponent of 3.

$$V_a^g = V_0^g \exp\left[-\left(\frac{AL}{\beta K_0}\right)^3\right]$$
(7.24)

$$A = -RT/\ln\left(\frac{P_0}{p}\right)$$
(7.25)

where

V_a^g is the adsorption quantity for the specific segment of the pore size distribution being considered at relative pressure P/P_0,

V_0^g is the total pore volume for the specific segment of the pore size distribution being considered,

K_0 is the product of the mean characteristic energy E_0 with the similarity coefficient β, and

L is the mean pore size of the specific segment of the pore size distribution being considered.

L values of all the pore groups considered are assumed to follow a Gamma-type distribution (i.e. essentially a two-metric continuous probability distribution, Carrott and Carrott 1999) expressed by Eq. (7.26):

$$\frac{dv}{dL} = \frac{3v_0 a^m L^{m-1} \exp(-aL^3)}{\Gamma(m)}$$
(7.26)

where

v_0 is the total micropore volume, and,

constants a and m are both related to the mean and width of the distribution.

Inserting Eq. (7.26) into Eq. (7.24) results in a Laplace transform and lead to Eq. (7.27), which is commonly referred to as the Stoeckli equation:

$$V_a = V_0 \left[\frac{a}{a + (A/\beta K_0)^3} \right]^m$$
(7.27)

where

V_a is the measured adsorption, and

V_0 is the total micropore volume.

It is not possible to determine the values of a and K_0 independently from Eq. (7.27). The method proposed by Carrott and Carrott (1999), and applied by Li et al. (2016a, b) to the Chongqing area shales, requires K_0 initially to be estimated using Eq. (7.24). Once a realistic estimate of K_0 is available the parameters a, m, and V_o can be derived from the fit of Eq. (7.27) to the measured data points. Once values for these variables are derived the pore size distribution for the sample tested can be generated from Eq. (7.26). From the description provided it is clear that the Stoeckli method is quite cumbersome mathematically to implement.

All of the isotherms described involve assumptions and are unlikely to provide exact and precise values for the pore volume, pore surface area or pore size distribution. Although this leads to uncertainty in the errors of the values derived, it does not significantly downgrade the usefulness of these methods in organic-rich shale porosity characterization. By running low-pressure adsorption tests (ideally using N_2 and CO_2) on a series of samples from a shale province displaying a range of TOC and levels of thermal maturity, and applying one or several of the isotherm methods described, the relative pore size distributions, pore volumes and total surface area values derived tend to be informative, meaningful and useful in characterizing the gas storage and/or production capabilities of each of those samples. The pore range over which adsorption analysis is generally considered to be viable and can be meaningfully applied is 0.25 nm < porosity < ~800 nm (2.5 Å < porosity < 8000 Å) (Meyer and Klobes 1999; Klobes et al. 2006).

7.3   Impact of Particle Crush-Size on Low-Pressure Gas Adsorption Studies

Ideally for conducting low-pressure gas analysis (LPGA) studies, the samples are crushed to finer-sizes as it provides a reduced the path-length and in doing so allows the gas to access the complex pore-structures present within a sample (Kuila and Prasad 2013). The Gas Research Institute (GRI) first introduced the crush-rock method for facilitating shale-core studies (Luffel and Guidry 1992). As there is no widely-accepted standard crush-size, different researchers have used different crush-sizes for LPGA experiments viz. 8 mm (Schmitt et al. 2013), 1.4 mm–250 microns (Mohammad et al. 2013), 1 mm–800 microns (Hazra et al. 2018a), <250 microns (most widely used; Ross and Bustin 2009; Strąpoć et al. 2010; Chalmers et al. 2012; Clarkson et al. 2013; Yang et al. 2014; Wang et al. 2016; Hazra et al. 2018b). Recently, several researchers have highlighted the impact of sample-sizes on LPGA results (Chen et al. 2015; Han et al. 2016; Mastalerz et al. 2017; Wei et al. 2016; Hazra et al. 2018c).

Chen et al. (2015) observed that by lowering the particle crush-size for gas adsorption experiments, mesopore volumes increased while and micropore volumes showed variable results. They observed that by reducing sample crush-sizes, the hysteresis loops became tighter due to better connectivity of pores and attained adsorption equilibrium early. Similarly, Han et al. (2016) in their study conducted on a Longmaxi shale (China) with crush-sizes ranging from 4 mm to 58 microns, observed increased pore volumes, surface areas, tightening of hysteresis loops with lowering of crush-sizes. Based on their results, they opined that particle-crush sizes for shales should be kept below 113 microns when conducting LPGA studies. Wei et al. (2016) on the other hand, based on their study on Sichuan Basin shales, suggested using samples with crush-sizes between 250 and 140 microns for LPGA studies. Mastalerz et al. (2017) based on their study on Illinois basin shales with distinct thermal maturity levels, proposed using 75 microns sample-sizes for LPGA experiments. Similar to the observations of earlier researchers, Hazra et al. (2018c) also observed significant variations in N_2 and CO_2 gas-adsorption results by changing sample crush-sizes from 1 mm to 53 microns, for two shales with different TOC contents and thermal maturity levels. They documented opening of earlier isolated/closed/constricted pores as sample crush-sizes reduced from 1 mm to 212 microns. However, with further reduction of crush size for the samples (212–75 microns), they observed alteration and destruction of some of the mesoporous and microporous structures. It is thus recommended that particle crush-sizes should be kept at around 212 microns.

Figure 7.6 displays the nitrogen gas adsorption isotherm of two shale samples with contrasting thermal maturity levels and TOC content, at two particle crush-sizes of 2–1 mm and 212–75 microns. Distinct pore structural properties can be observed for the samples at the two crush-sizes and are listed in Table 7.2. For both the samples, with decreasing crush-sizes, the volume of gas adsorbed and the BET SSA increase.

For the high-TOC shale (BMF2), with a large particle crush-size (2–1 mm; Fig. 7.6a1), the nitrogen adsorption-desorption isotherms show distinct hysteresis, indicating the presence of mesopores (Sing et al. 1985). This isotherm resembles, the shape of the type IV isotherm (Fig. 7.1), but lacks the presence of a plateau at high relative pressures, typical of type IV isotherms. Neither do the observed isotherms show any steep slopes in the relative pressure range of 0.98–1.00 typical of type II macroporous materials. Isotherm curves of this shape observed in shales have also previously been classified as Type IIB curves by Rouquerol et al. (1998), and Kuila and Prasad (2013). At 212–75 microns particle crush-size (Fig. 7.6a2), the slope in the relative pressure range of 0.98–1.00 is steeper, indicating more macropores.

A new parameter ΔV_G i.e. the difference in the volumes of gas adsorbed in the last two pressure steps introduced by Hazra et al. (2018a), shows higher values for the 212–75 microns sized sample-split compared to the coarser sized sample-split (Table 7.2). The 212–75 microns sample-split also shows a higher gas adsorbed volume. Further, the hysteresis pattern is tighter/narrower for the 212–75 microns

Table 7.2 Low-pressure N_2 gas adsorption characteristics of two Permian shales from India at two particle crush-sizes

Sample details	N2-GA parameters	2–1 mm	212–75 microns
S.N.: **BMF2** (High-TOC) Barren measures formation Raniganj basin TOC: 7.84 wt% T_{max}: 439 °C	BET SSA (m²/g)	18.03	19.77
	Parameter C from the BET equation	489.015	499.932
	Average pore radius (Å)	23.72	27.74
	BJH pore volume (cc/g)	0.022	0.026
	V_G (cc/g)	14.59	17.73
	ΔV_G (cc/g)	1.08	2.52
S.N.: **Bar2** (Carbonaceous shale) Barakar formation Jharia basin TOC: 23.18 wt% T_{max}: 477 °C	BET SSA (m²/g)	2.26	5.86
	Parameter C from the BET equation	−132.019	221.765
	Average pore radius (Å)	25.48	34.57
	BJH pore volume (cc/g)	0.003	0.009
	V_G (cc/g)	1.86	6.55
	ΔV_G (cc/g)		

Note V_G represents the maximum volume of gas (cc/g) adsorbed by the sample. ΔV_G represents the difference in the volumes of gas adsorbed at the last two pressure steps. The greater the difference, the steeper would be the slope of the isotherm at P/P_0 close to 1, and thereby would indicate presence of macroporous structures

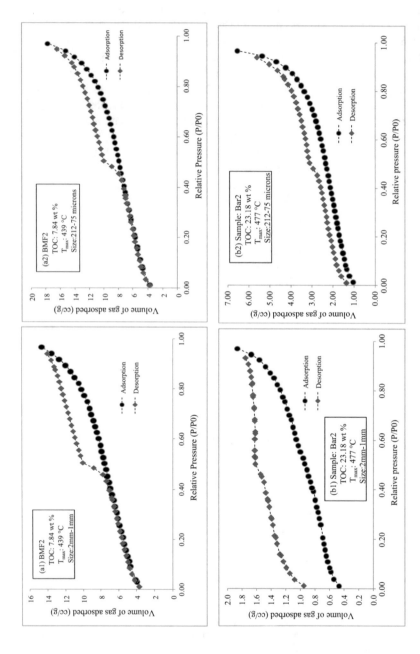

Fig. 7.6 Nitrogen gas adsorption-desorption isotherms of the high-TOC early mature Barren Measures shale (**a1** and **a2**) and the overmature carbonaceous shale sample (**b1** and **b2**) at two contracting particle crush-sizes. The blue circles represent the adsorption isotherm and the red diamonds depict the desorption isotherm which taken together define the hysteresis pattern

sample-split, compared to the coarser size-split. This tightness of the hysteresis pattern is interpreted to be due to the lesser distance or travel-path needed for desorption in finer-sized sample-splits, than the coarser-sized sample-splits.

The increased volume of gas adsorbed, greater BET SSA, tightening of the hysteresis loop, all imply enhanced accessibility of porous-structures by nitrogen injected into finer-sized sample-splits. In contrast, assess to pore structures is likely to be partially restricted by the shale matrix at the coarser-sized sample-splits. On the other hand, the average pore size and BHJ pore volume, increased from the 2–1 mm sized-split to 212–75 microns sized-split (Table 7.2). These results suggest that with fining particle crush-sizes, earlier isolated or constricted or blind porous structures (both mesopores and macropores) become more accessible for the injected nitrogen at finer crush-sizes. Hazra et al. (2018c) noted a decrease in BET SSA, jump in pore sizes and pore volumes when samples were crushed to even smaller particle sizes (<75 microns), indicating liberation of porous structures and also possible alteration of mesoporous structures by the more intensive sample crushing involved.

For the over mature carbonaceous shale sample (Bar2), the impact of particle crush-size on nitrogen gas adsorption measurements was even more distinctive. For the coarser size-split, the BET SSA is unreliable, as the C parameter (used in the BET equation) is negative (Table 7.2). When C is too low (<2) or negative, the BET method is considered unreliable (Thommes et al. 2015). While for the 2–1 mm sized sample-split, the desorption branch is observed to be open and never merges with the adsorption branch (Fig. 7.6b1), for the finer sized-split, the hysteresis pattern is tighter/narrower and closer, although displaying some hysteresis, indicating the presence of mesopores (Fig. 7.6b2). The increase in BJH pore volume, volume of gas adsorbed for the 212–75 microns size-split suggest an opening-up or further exposure of pore spaces which were constricted/restricted in the coarser sample fraction.

The results for LP-CO_2-GA for the two samples BMF2 and Bar2, across two particle crush-sizes are listed in Table 7.3. Figure 7.7 shows the CO_2 gas adsorption isotherms for the two samples at the two-tested particle crush-sizes. With decreasing particle crush-sizes, for both the samples, the gas adsorption capacity, Dubinin-Astakhov (D-A) micropore surface area (MSA), micropore volume (MV), Dubinin-Radushkevich (D-R) MSA were observed to increase. This implies that micropores which are constricted or isolated at coarser crush-sizes become liberated when the samples are crushed to finer sizes, and thus become more accessible to the CO_2 gas. Furthermore, the MSA and MV were observed to be higher in the over-mature shale Bar2, while and pore size was observed to be lower in shale Bar2 relative to the early mature shale BMF2. These LP-CO_2-GA results for samples Bar2 and BMF2 are contradictory to the results obtained from LP-N_2-GA conducted on those samples. An explanation for these results is provided in Sects. 7.4 and 7.5.

Table 7.3 Low-pressure CO_2 gas adsorption characteristics of two India shales at two particle crush-sizes

Sample details	CO_2-GA parameters	2–1 mm	212–75 microns
S.N.: **BMF2** (High-TOC) Barren measures formation Raniganj basin TOC: 7.84 wt% T_{max}: 439 °C	D-A MSA	21.75	31.91
	DA MV	0.009	0.013
	DA mean equivalent pore radius	8.40	8.30
	D-R MSA	21.99	33.04
	V_G (cc/g)		
S.N.: **Bar2** (Carbonaceous shale) Barakar formation Jharia basin TOC: 23.18 wt% T_{max}: 477 °C	D-A MSA	39.52	47.89
	DA MV	0.017	0.020
	DA mean equivalent pore radius	8.35	8.20
	D-R MSA	41.28	49.01
	V_G (cc/g)	4.28	5.57

Note D-A: Dubinin-Astakhov; D-R: Dubinin-Radushkevich; MSA: Micropore surface area; MV: Micropore volume; VG: Volume of gas adsorbed (cc/g)

7.4   Controlling Parameters for Pore Structural Features in Shales

Whether the organic-fraction or the inorganic-fraction of shales controls their SSA continues to be a point of debate. While some studies report a positive correlation between TOC and BET SSA of shales (Fig. 7.8), other studies have reported either no correlation or a negative correlation between these two variables (Fig. 7.9). Additionally, some studies have addressed the impact of thermal maturity levels on specific surface area of shales. Ross and Bustin (2009) observed higher BET SSA for thermally mature Devonian-Mississippian shales compared to thermally immature Jurassic shales from the Western Canadian Sedimentary Basin (Fig. 7.9). They further noted that the Jurassic shales were characteristically marked by the presence of amorphous bituminite in the pore space, which might have reduced the BET SSA measured for such shale samples. Similarly, Hazra et al. (2018b) observed higher BET SSA values for thermally impacted (i.e., locally heated by igneous intrusions inducing contact metamorphism) Permian shales from Raniganj basin, irrespective of their TOC contents. Compared to the non-thermally impacted shales, the thermally metamorphosed ones showed higher SSA, pore volumes, and lower pore sizes. These findings suggest that organic-rich shales marked by higher thermal maturities are likely to be characterized by higher surface areas. This increase in SSA is considered to be caused due by the expulsion of petroleum from the organic-matter and concomitant formation of secondary porosity.

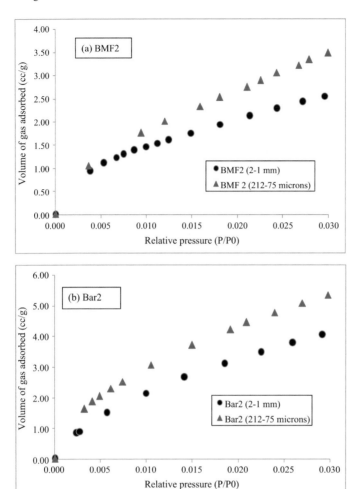

Fig. 7.7 Low-pressure CO_2 gas adsorption isotherms for Indian Permian shale samples BMF2 (**a**) and Bar2 (**b**)

Chen and Xiao (2014) artificially matured two organic-rich Upper Permian shales and one organic-lean Oligocene shale from China. For the organic-rich samples, they observed an increase in micropore volumes, micropore and mesopore surface areas from the oil-window level of maturity to post mature levels of thermal maturity (i.e., vitrinite reflectance of about 3.50%). Beyond that level of thermal maturity (i.e. extremely high levels of thermal maturity) slight decreases in micropore surface areas were recorded for these samples. However, the mesopore volumes for these shales were observed to increase progressively with increasing thermal maturity. These findings suggest that the increase in nanoporosity up to equivalent vitrinite reflectance of 3.5% is likely to be caused by the creation of nanopores within the

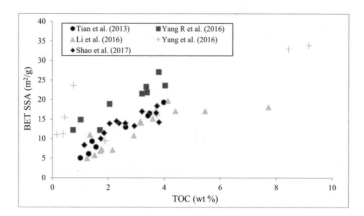

Fig. 7.8 Positive correlation between BET surface area and TOC content of shales. Data compiled from Tian et al. (2013), Li et al. (2016a, b), Yang et al. (2016a, b), and Shao et al. (2017)

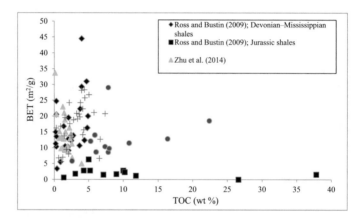

Fig. 7.9 Lack of a distinctive correlation between BET surface area and TOC content of shales. Data compiled from Ross and Bustin (2009), Zhu et al. (2014), Xia et al. (2017), and Hazra et al. (2018a)

organic-matter as thermal maturity increases to that point. However, at extremely high levels of thermal maturity (Ro > 3.5%) some destruction of nanoporosity ensues.

Although an overall increase in nanoporosity, micropore and mesopore volumes and surface areas with increasing thermal maturity levels is to be expected in organic rich formations, the lack of a positive correlation between these physical rock properties and the TOC content observed in some studies is counterintuitive. Comparing BET SSA of Permian coals and shales from different Indian basins reveals that for coal samples, the BET SSA is much lower compared to those of shales (Fig. 7.10). All these Permian shale and coal samples from India are of nearly similar thermal maturity levels (primarily within the oil-window and early condensate wet-gas window based on vitrinite reflectance measurements) and are predominantly com-

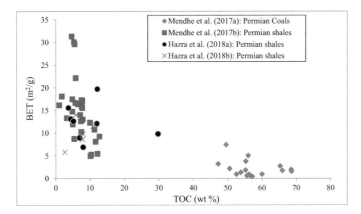

Fig. 7.10 BET SSA versus TOC for Permian shales and coals from India. Data compiled from Mendhe et al. (2017a, b) and Hazra et al. (2018a, b)

posed of type III-IV organic-matter (Mendhe et al. 2017a, b; Hazra et al. 2018a, b). Figure 7.10 also reveals a distinctive decreasing trend in BET SSA from shales to coals. This decreasing nitrogen sorption capacity and BET SSA for the Permian shales and coals with increasing TOC, based on nitrogen sorption measurements, at first inspection suggests that the magnitude of organic matter content in these formations is not related to the evolution of SSA as thermal maturity progresses. However, such a conclusion would contradict the observation that much of the gas in shale-gas and coal bed methane reservoirs is widely reported to exist in the sorbed state, primarily within the organic-matter and secondarily within the mineral matter (Cheng and Huang 2004; Chalmers and Bustin 2007; Varma et al. 2014a, b). An alternative explanation for both shales and coals could be the partial inability of nitrogen to fully access and penetrate the complex porous structure within organic-matter, particular at the nanopore scale. The presence of bituminite or condensate oil within certain pores of organic-matter in some formations may hinder the accessibility of porous structures by the nitrogen gas for immature and early-mature samples of oil-prone shales (Ross and Bustin 2009). In such case, as thermal maturity increases, prompting the expulsion of some of the petroleum more pores would become accessible to nitrogen injection. However, this explanation is an unlikely one for shales with predominantly type III and type IV kerogen which generate predominantly natural gas.

For the two shales BMF2 and Bar2 (Tables 7.2 and 7.3) at the 212 microns sample-split, the results of N_2 and CO_2 gas adsorption results can be observed to be contradictory to one another. For the high-TOC Barren Measures Formation shale (BMF2), using N_2 as adsorbate, the SSA and average pore radius are revealed to be greater and smaller than the more-carbon-rich shale (Bar2), respectively. However, when CO_2 is used as the adsorbate, the carbonaceous shale (Bar2) showed higher D-R and D-A surface areas and pore volumes, and lower D-A pore radius compared to the high-TOC shale (BMF2). These contradictory results raise some doubts about the way in which low-pressure gas adsorption data is interpreted for organic-rich formations. The lack of a correlation between TOC and BET SSA for nitrogen gas adsorption measurements in these samples supports the view that a lack of pene-trability of injected nitrogen into the complex pore network in the organic matter contained in some shales at certain levels of maturity.

At the very low experimental temperatures involved (about −197.3 °C), the ther-mal energy of nitrogen may be insufficient to access complex porosity systems involv-ing extensive and isolated nanoporosity. On the other hand, low pressure carbon dioxide gas adsorption measurements that are conducted at 0 °C, involve gas injec-tion at higher thermal energies which may be more effective at penetrating through narrow and complex connecting pathways at least partially connecting the prevailing nanoporosity network. This implies that carbon dioxide gas adsorption measurements provide more representative measurements of the micropore surface area (Unsworth et al. 1989; Ross and Bustin 2009) and that nitrogen gas adsorption measurements should be considered as prone to underestimate micropore surface area.

For shale samples in which the organic-matter displays inherently more meso-porous structures relative to nanoporous structures the nitrogen adsorption measure-ments may provide a realistic approximation to BET SSA. On the other hand, for those shales displaying more nanoporosity and complex pore networks and pore-size distributions nitrogen adsorption measurements should be questioned or considered as minimum estimates of the prevailing SSA. This distinction between the effective-ness of nitrogen and carbon dioxide gas adsorption measurements for organic-rich formations has significant implications for estimates of the gas storage potential of such reservoirs. it suggests that nitrogen adsorption measurements should not be relied upon in isolation and that more confidence should be attributed to carbon dioxide gas adsorption measurements in evaluations of the reservoir potential of organic-rich shales.

7.5 Fractal Dimensions in Shales

Determining the presence of fractal geometries within porous materials/reservoir rocks using gas adsorption studies is a well-established process (Mandelbrot 1975; Pfeifer and Avnir 1983). The theory of fractal geometry was first presented by Mandelbrot (1975) to depict asymmetric and fragmented systems without any dis-tinctive length-scale. It relies mainly on the concept of self-similarity (Mahamud and Novo 2008). Information regarding different pore parameters viz. pore volumes, SSA, pore sizes may be variable depending on the methodology adopted and the mod-els used to calculate them. However, they do not necessarily give detailed information about the fabric, shape and texture of the pore-surface or pore structures present. The fractal dimensions of porous materials on the other hand does not depend on the pore size or the amount of porous structures present but reflect fundamental traits of the surface itself (Mahamud and Novo 2008). Because their governing rules simplify complex features in nature, fractals have found application in relating the pore scale understandings to the core scale features such as permeability and flowage (Liu et al. 2014; Sakhaee-Pour and Li 2016) and the microscopic fabric of tight formations.

Fractal dimension D is the most critical component in fractal geometry. D evaluates the surface-irregularities and structural-irregularities of a solid substance (Jaroniec 1995). D values vary between 2 (perfectly smooth surface of a solid) to 3 (maximum geometrical-complexity on the surface of a solid). Fractal dimensions of organic-sedimentary rocks estimated using low-pressure gas adsorption technique helps in evaluating and quantifying their pore surface and structural complexities, which controls their adsorption and desorption behaviors (Mahamud and Novo 2008; Yao et al. 2008; Javadpour 2009; Cai et al. 2013; Clarkson et al. 2013; Yuan et al. 2014; Yang et al. 2014; Wang et al. 2016; Wood and Hazra 2017). Further, it has also been established that nitrogen, as an adsorbate, provides a more reliable surface irregularity-measure and fractal geometry, compared to other gases such as krypton or argon (Ismail and Pfeifer 1994).

Although several methods exist to calculate the fractal dimension of coals and shales using low-pressure nitrogen adsorption data, the Frenkel-Halsey-Hill adsorption isotherm equation (FHH model) (Yang et al. 2014; Hu et al. 2016; Bu et al. 2015; Li et al. 2016a, b) is the most commonly applied. The FHH model applies Eq. (7.28) (Qi et al. 2002; Yao et al. 2008):

$$\ln\left(\frac{V}{V_0}\right) = A\left[\ln\left(\ln\left(\frac{P_0}{P}\right)\right)\right] + \text{constant} \qquad (7.28)$$

where

P is the equilibrium pressure,
P_0 is the saturation pressure of the gas,
V is the volume of adsorbed gas molecules,
V_0 is the volume of monolayer coverage, and,
A is the power law exponent which depends on the fractal dimension (D) and the mechanisms of adsorption.

D can be calculated from the slope (S) of the straight line in the $\ln V$ versus $\ln[\ln(P_0/P)]$ FHH plot using Eq. (7.29) or (7.30) (Qi et al. 2002; Rigby 2005).

$$S = D - 3 \qquad (7.29)$$

$$S = (D - 3)/3 \qquad (7.30)$$

It is well established that at low relative-pressures during LP-N$_2$-GA, the adsorption process is controlled by van der Waals attraction forces i.e., when the liquid/gas surface tension forces are insignificant (Jaroniec 1995). In such circumstances, the van der Waals forces at the gas/solid interface form the adsorbed film which replicates the surface roughness. The fractal dimension existent at such lower relative pressures (0–0.5) is assigned as fractal dimension $D1$, which is referred as the *pore-surface-fractal dimension* (Khalili et al. 2000; Yao et al. 2008). It was initially suggested that the fractal dimension in such circumstances (where adsorption is controlled by van der Waals dominated forces) should be calculated using Eq. (7.30) (Jaroniec 1995). On the other hand, Eq. (7.29) has been traditionally used for calculating the fractal dimension at higher relative pressures (0.5–1.0), where the interface is controlled by the liquid/gas surface tension i.e. capillary condensation/evaporation (Pfeifer et al. 1989; Jaroniec 1995). The fractal dimension characteristic of such higher relative pressures (0.5–1.0) is denoted as $D2$, which is referred to as the *pore-structural-fractal dimension* (Khalili et al. 2000; Yao et al. 2008).

However, fractal dimension $D1$ calculated using Eq. (7.29) has seldom been found to be lower than 2, which is a strong deviation from the range of fractal values (2–3) observed for porous materials (Pfeifer and Avnir 1983; Xie 1996). Figure 7.11 shows a comparative plot of $D1$ values for ninety shale samples from different geological settings/countries, obtained using Eqs. (7.29) and (7.30). Figure 7.11 reveals that when Eq. (7.30) is used to calculate $D1$ (y-axis), the values are mostly <2, and hence are unreliable. On the other hand, when $D1$ is calculated using Eq. (7.29), the values vary between 2 and 3 (x-axis in Fig. 7.11). Under such circumstances, although $D1$ may be evaluated using Eq. (7.30) (Ismail and Pfeifer 1994), several authors have preferred to apply Eq. (7.29) to calculate $D1$ because it provides more meaningful results (Bu et al. 2015; Shao et al. 2017; Hazra et al. 2018a, b).

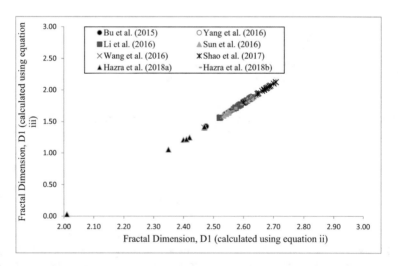

Fig. 7.11 Cross-plot showing the variation of D1 values calculated using Eqs. (7.29) and (7.30). When D1 is calculated using Eq. (7.30), the values are mostly below 2, while using Eq. (7.29) generates values between 2 and 3. The data presented is compiled from Bu et al. (2015), Li et al. (2016a, b), Wang et al. (2016), Yang et al. (2016a), Sun et al. (2016), Shao et al. (2017), and Hazra et al. (2018a, b)

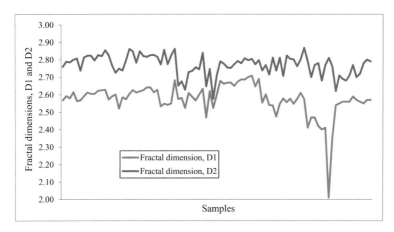

Fig. 7.12 Plot showing the variation of D1 and D2 values calculated using Eq. (7.29). In all the samples D2 is greater than D1. The data sources are the same as for Fig. 7.11

Figure 7.12 displays $D1$ versus $D2$ calculated using Eq. 7.29 for the same ninety shales considered in Fig. 7.11. $D2$ values are always greater than $D1$ values. Similarly, for the two samples whose properties are listed in Tables 7.2 and 7.3, the fractal dimension $D2$ is greater than fractal dimension $D1$ (Table 7.4; Fig. 7.13). It is well documented that for pores of molecular dimensions (ultra-micropore), the pore-filling occurs at the initial phases of adsorption i.e. at very low P/P_0 (Rouquerol et al. 1998). For the narrowest micropores, the pore-walls occur close to one another, which increase the interaction-energy between the adsorbate and the adsorbent, leading to micropore filling at very low initial P/P_0 (Rouquerol et al. 1998; Lowell et al. 2004). The fractal dimension $D1$ (pore-surface fractal) is calculated at low relative pressure (P/P_0 less than 0.5). Since nitrogen at low experimental temperatures does not have sufficient thermal energy to penetrate through the narrowest and most constricted pores (i.e., those in the micropore range), the representation of $D1$ is always likely to be more poorly defined or undercounted compared to the pore-surface complexity actually present. This is especially so for the complex micropore size distributions characteristic of organic-rich shales. The fractal properties for the two shales BMF2 and Bar2 shed further light on this point.

For both shale samples, BMF2 and Bar2, fractal dimension D2 is greater than D1. Although sample Bar2 is marked by a greater concentration of micropores and higher micropore surface area compared to sample BMF2 (as revealed by LP-CO_2-GA; Table 7.3), its fractal dimension $D1$ is smaller than that of sample BMF2, although sample Bar2 is more mature (postmature; $T_{max} = 477\ °C$) than sample BMF2 (early mature; $T_{max} = 439\ °C$). It is now well established that with increasing thermal maturity levels, especially beyond the oil-window, secondary microporosity is created within organic-matter as petroleum is liberated from the system (Behar

Table 7.4 Fractal fitting parameters and fractal dimensions calculated using the FHH model for 2Permian shale samples (India)

Sample number	P/P_0 (0.01–0.50)				P/P_0 (0.50–1.00)			
	S_1	R_1^2	$D1$		S_2	R_2^2	$D2$	
			$3+S$	$3+3S$			$3+S$	$3+3S$
BMF2	−0.414	0.999	2.59	1.76	−0.265	0.998	2.74	2.21
Bar2	−0.437	0.994	2.56	1.69	−0.336	0.998	76	1.99

Note S_1, R_1^2, and D1 represent the slope of the straight line, coefficient of determination, and fractal dimension, respectively, in the lnV versus ln [ln(P$_0$/P)] FHH plot for the relative pressure range (P/P$_0$) of 0.01–0.50 (Fig. 7.13). S_2, R_2^2, and D2 represent the slope of the straight line, coefficient of determination, and fractal dimension respectively, in the lnV versus ln [ln(P$_0$/P)] FHH plot for the relative pressure range (P/P$_0$) of 0.50–1.00 (Fig. 7.13). Other LPGA properties of these two shales are mentioned in Tables 7.2 and 7.3. The fractal dimensions presented here were measured on a 212–75 microns sample-split

and Vandenbrouke 1987; Pommer and Milliken 2015). The less-complex measured pore-surface fractal dimension ($D1$) for the over-mature shale Bar2 with higher concentration of micropores and greater micropore surface area, relative to the early mature shale BMF2, is consistent with the undercounting (incomplete detection) of the microporosity by nitrogen adsorption analysis mentioned. This implies that fractal dimension $D1$ may be an underestimated, especially in the presence of complex porous-structures for thermally mature organic-rich shales.

In general, capillary condensation controls the filling of mesopores and macropores (i.e. pore-width > 2 nm) in porous shales. Before capillary condensation commences, multi-layer adsorption takes place on the pore walls (Sing 2001). Sun et al. (2016) based on this principle opined that when more gas molecules are adsorbed and more gas molecules cover the pore structure, the value of fractal dimension in this zone (i.e. $D2$) is likely to be higher than the fractal dimension in the zone where monolayer-multilayer adsorption takes place (i.e. $D1$). The data presented here reflect that relationship.

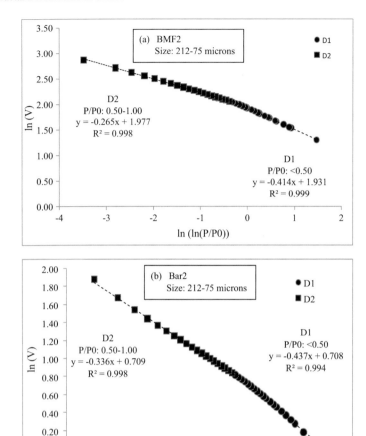

Fig. 7.13 lnV versus ln(ln(P₀/P)) from the nitrogen adsorption isotherms using the FHH model for the two samples BMF2 (**a**), and Bar2 (**b**) at 212–75 microns particle crush-size

References

Barrett EP, Joyner LG, Halenda PP (1951) The determination of pore volume and area distributions in porous substances. I. Computations from nitrogen isotherms. J Am Chem Soc 73(1):373–380

Behar F, Vandenbroucke M (1987) Chemical modelling of kerogens. Org Geochem 11:15–24

Bernard S, Wirth R, Schreiber A, Schulz H-M, Horsfield B (2012a) Formation of nanoporous pyrobitumen residues during maturation of the Barnett Shale (Fort Worth Basin). Int J Coal Geol 103:3–11

Bernard S, Horsfield B, Schulz H-M, Wirth R, Schreiber A (2012b) Geochemical evolution of organic-rich shales with increasing maturity: a STXM and TEM study of the Posidonia Shale (Lower Toarcian, northern Germany). Mar Pet Geol 31:70–89

Brunauer S, Deming LS, Deming WS, Teller E (1940) On a theory of the van der Waals adsorption of gases. J Am Chem Soc 62:1723–1732

Bu H, Ju Y, Tan J, Wang G, Li X (2015) Fractal characteristics of pores in nonmarine shales from the Huainan coalfield, eastern China. J Nat Gas Sci Eng 24:166–177

Cai Y, Liu D, Pan Z, Yao Y, Li J, Qiu Y (2013) Pore structure and its impact on CH_4 adsorption capacity and flow capability of bituminous and subbituminous coals from Northeast China. Fuel 103:258–268

Camp WK, Wawak B (2013) Enhancing SEM grayscale images through pseudocolor conversion: examples from Eagle Ford, Haynesville, and Marcellus Shales. In: Camp WK, Diaz E, Wawak B (eds) Electron Microscopy of Shale Hydrocarbon Reservoirs 102. AAPG Memoir, pp 15–26

Cardott BJ, Landis CR, Curtis ME (2015) Post-oil solid bitumen network in the Woodford Shale, USA—a potential primary migration pathway. Int J Coal Geol 139:106–113

Carrott P, Carrott MR (1999) Evaluation of the Stoeckli method for the estimation of micropore size distributions of activated charcoal cloths. Carbon 37(4):647–656

Chalmers GRL, Bustin RM (2007) The organic matter distribution and methane capacity of the Lower Cretaceous strata of Northeastern British Columbia, Canada. Int J Coal Geol 70:223–239

Chalmers GR, Bustin RM, Power IM (2012) Characterization of gas shale pore systems by porosimetry, pycnometry, surface area, and field emission scanning electron microscopy/transmission electron microscopy image analyses: examples from the Barnett, Woodford, Haynesville, Marcellus, and Doig units. AAPG Bull 96:1099–1119

Chen Y, Wei Y, Mastalerz M, Schimmelmann A (2015) The effect of analytical particle size on gas adsorption porosimetry of shale. Int J Coal Geol 138:103–112

Chen J, Xiao X (2014) Evolution of nanoporosity in organic-rich shales during thermal maturation. Fuel 129:173–181

Chen Z, Jiang C (2016) A revised method for organic porosity estimation in shale reservoirs using Rock-Eval data: example from the Duvernay Formation in the Western Canada Sedimentary Basin. AAPG Bull 100:405–422

Cheng AL, Huang WL (2004) Selective adsorption of hydrocarbon gases on clays and organic matter. Org Geochem 35:413–423

Clarkson CR, Bustin RM (1999) The effect of pore structure and gas pressure upon the transport properties of coal: a laboratory and modeling study. 1. Isotherms and pore volume distributions. Fuel 78(11):1333–1344

Clarkson CR, Solano N, Bustin RM, Bustin AMM, Chalmers GRL, Hec L, Melnichenko YB, Radlinskid AP, Blachd TP (2013) Pore structure characterization of North American shale gas reservoirs using USANS/SANS, gas adsorption, and mercury intrusion. Fuel 103:606–616

Curtis JB (2002) Fractured shale-gas systems. AAPG Bull 86:1921–1938

Curtis ME, Ambrose RJ, Sondergeld CH, Rai CS (2010) Structural characterization of gas shales on the micro- and nano-scales. SPE-137693, CSUG/SPE Canadian Unconventional Resources and International Petroleum Conference. October 19–21, 2010, Calgary, Alberta

Curtis ME, Cardott BJ, Sondergeld CH, Rai CS (2012) Development of organic porosity in the Woodford Shale with increasing thermal maturity. Int J Coal Geol 103:26–31

Desbois G, Urai JL, Kukla PA (2009) Morphology of the pore space in clay stones—evidence from BIB/FIB ion beam sectioning and cryo-SEM observations. eEarth Discussions 4:1–19

Dubinin MM, Astakhov VA (1971) Description of adsorption equilibria of vapors on zeolites over wide ranges of temperature and pressure. Adv Chem 102(69):65–69. https://doi.org/10.1021/ba-1971-0102.ch044

Dubinin MM, Radushkevich LV (1947) Equation of the characteristic curve of activated charcoal. Proc Acad Sci USSR 55:331–333

Gregg SJ, Sing KSW (1982) Adsorption, surface area, and porosity, 2nd edn. Academic Press, New York

Halsey G (1948) Physical adsorption on non-uniform surfaces. J Chem Phys 16:931

Han H, Cao Y, Chen S-J, Lu J-G, Huang C-X, Zhu HH, Zhan P, Gao Y (2016) Influence of particle size on gas-adsorption experiments of shales: an example from a Longmaxi Shale sample from the Sichuan Basin, China. Fuel 186:750–757

Harkins WD, Jura G (1944) Surface of solids. XIII: a vapor adsorption method for the determination of the area of a solid without the assumption of a molecular area, and the areas occupied by nitrogen and other molecules on the surface of a solid. J Am Chem Soc 66(8):1366–1373

Hazra B, Wood DA, Vishal V, Varma AK, Sakha D, Singh AK (2018a) Porosity controls and fractal disposition of organic-rich Permian shales using low-pressure adsorption techniques. Fuel 220:837–848

Hazra B, Wood DA, Kumar S, Saha S, Dutta S, Kumari P, Singh AK (2018b) Fractal disposition and porosity characterization of Lower Permian Raniganj Basin Shales, India. J Nat Gas Sci Eng 59:452–465

Hazra B, Wood DA, Vishal V, Singh AK (2018c) Pore-characteristics of distinct thermally mature shales: influence of particle sizes on low pressure CO_2 and N_2 adsorption. Energy Fuels 32(8):8175–8186

Hill TL (1952) Theory of physical adsorption. Adv Catal IV:211–257

Hu J, Tang S, Zhang S (2016) Investigation of pore structure and fractal characteristics of the lower Silurian Longmaxi shales in western Hunan and Hubei provinces in China. J Nat Gas Sci Eng 28:522–535

Ismail IMK, Pfeifer P (1994) Fractal analysis and surface roughness of nonporous carbon fibers and carbon blacks. Langmuir 10:1532–1538

Jaroniec M (1995) Evaluation of the fractal dimension from a single adsorption isotherm. Langmuir 11:2316–2317

Jarvie DM, Hill RJ, Ruble TE, Pollastro RM (2007) Unconventional shale-gas systems: the Mississippian Barnett Shale of north-central Texas as one model for thermogenic shale-gas assessment. AAPG Bull 91(4):475–500

Javadpour F (2009) Nanopores and apparent permeability of gas flow in mudrocks (shales and siltstone). J Can Pet Technol 48(8):16–21

Jennings DS, Antia J (2013) Petrographic characterization of the Eagle Ford Shale, south Texas: mineralogy, common constituents, and distribution of nanometer-scale pore types. In: Camp W, Diaz E, Wawak B (eds) Electron microscopy of shale hydrocarbon reservoirs, vol 102. AAPG Memoir, pp 101–113

Kang SM, Fathi E, Ambrose RJ, Akkutlu IY, Sigal RF (2011) Carbon dioxide storage capacity of organic-rich shales. Soc Pet Eng J 16(4):842–855

Khalili NR, Pan M, Sandí G (2000) Determination of fractal dimension of solid carbons from gas and liquid phase adsorption isotherms. Carbon 38:573–588

Klobes P, Meyer K, Munro RG (2006) Surface area measurements for solid materials. National Institute of Standards and Technology (NIST) Recommended Practice Guide Special Publication 960–17. 89 pages

Kuila U, Prasad M (2013) Specific surface area and pore-size distribution in clays and shales. Geophys Prospect 61:341–362

Langmuir I (1918) The adsorption of gases on plane surfaces of glass, mica and platinum. J Am Chem Soc 40(1918):1361

Leddy N (2012) Surface area and porosity. CMA Analytical workshop. https://www.tcd.ie/CMA/misc/Surface_area_and_porosity.pdf

Li T, Tian H, Chen J, Cheng L (2016a) Application of low pressure gas adsorption to the characterization of poresize distribution of shales: an example from Southeastern Chongqing area, China. J Nat Gas Geosci 1:221–230

Li A, Ding W, He J, Dai P, Yin S, Xie F (2016b) Investigation of pore structure and fractal characteristics of organic-rich shale reservoir: a case study of Lower Cambrian Qiongzhusi formation in Malong block of eastern Yunnan Province, South China. Mar Pet Geol 70:46–57

Liu T, Zhang XN, Li Z, Chen ZQ (2014) Research on the homogeneity of asphalt pavement quality using X-ray computed tomography (CT) and fractal theory. Constr Build Mater 68:587–598

Liu B, Schieber J, Mastalerz M (2017) Combined SEM and reflected light petrography of organic matter in the New Albany Shale (Devonian-Mississippian) in the Illinois Basin: A perspective on organic pore development with thermal maturation. Int J Coal Geol 184:57–72

Löhr SC, Baruch ET, Hall PA, Kennedy MJ (2015) Is organic pore development in gas shales influenced by the primary porosity and structure of thermally immature organic matter? Org Geochem 87:119–132

Loucks RG, Reed RM, Ruppel SC, Jarvie DM (2009) Morphology, genesis, and distribution of nanometer-scale pores in siliceous mudstones of the Mississippian Barnett Shale. J Sediment Res 79:848–861

Lowell S, Shields JE, Thomas MA, Thommes M (2004) Characterization of porous Solids and powders: surface area, pore size and density. Springer Science. ISBN 978-90-481-6633-6

Luffel DL, Guidry FK (1992) New core analysis methods for measuring reservoir rock properties of Devonian shale. J Petrol Technol 44:1184–1190

Mandelbrot BB (1975) Les Objects Fractals: Forme, Hasard et Dimension. Flammarion, Paris

Mahamud MM, Novo MF (2008) The use of fractal analysis in the textural characterization of coals. Fuel 87:222–231

Mastalerz M, He L, Melnichenko YB, Rupp JA (2012) Porosity of coal and shale: insights from gas adsorption and SANS/USANS techniques. Energy Fuels 26:5109–5120

Mastalerz M, Schimmelmann A, Drobniak A, Chen Y (2013) Porosity of Devonian and Mississippian New Albany Shale across a maturation gradient: insights from organic petrology, gas adsorption, and mercury intrusion. AAPG Bull 97:1621–1643

Mastalerz M, Hampton L, Drobniak A, Loope H (2017) Significance of analytical particle size in low-pressure N_2 and CO_2 adsorption of coal and shale. Int J Coal Geol 178:122–131

Meyer K, Klobes P (1999) Comparison between different presentations of pore size distribution in porous materials. Fresenius J Anal Chem 363(2):174–178

Mendhe VA, Mishra S, Varma AK, Kamble AD, Bannerjee M, Sutay T (2017a) Gas reservoir characteristics of the Lower Gondwana Shales in Raniganj Basin of Eastern India. J Petrol Sci Eng 149:649–664

Mendhe VA, Bannerjee M, Varma AK, Kamble AD, Mishra S, Singh BD (2017b) Fractal and pore dispositions of coal seams with significance to coal bed methane plays of East Bokaro, Jharkhand, India. J Nat Gas Sci Eng 38:412–433

Milliken KL, Rudnicki M, Awwiller DN, Zhang T (2013) Organic matter-hosted pore system, Marcellus Formation (Devonian), Pennsylvania. AAPG Bull 97:177–200

Milner M, McLin R, Petriello J (2010) Imaging texture and porosity in mudstones and shales: comparison of secondary and ion-milled backscatter SEM methods. SPE-138975, CSUG/SPE Canadian Unconventional Resources and International Petroleum Conference, Calgary. October 19–21, 2010, Alberta, Canada

Mohammad ML, Rezaee R, Saeedi A, Al Hinai A (2013) Evaluation of pore size spectrum of gas shale reservoirs using low pressure nitrogen adsorption, gas expansion and mercury porosimetry: a case study from the Perth and Canning basins, Western Australia. J Petrol Sci Eng 112:7–16

Monson PA (2012) Understanding adsorption/desorption hysteresis for fluids in mesoporous materials using simple molecular models and classical density functional theory. Microporous Mesoporous Mater 160:47

Pfeifer P, Avnir D (1983) Chemistry nonintegral dimensions between two and three. J Phys Chem 79:3369–3558

Pfeifer P, Wu Y, Cole M, Krim J (1989) Multilayer adsorption on a fractally rough surface. Phys Rev Lett 62:1997

Pirngruber G (2016) Physisorption and pore size analysis. Characterization of Porous Solids-Characterization of catalysts and surfaces. Institut Francais du Petrole. 68 pages. https://www.ethz.ch/content/dam/ethz/special-interest/chab/icb/van-bokhoven-group-dam/coursework/Characterization-Techniques/2016/physisorption-pore-size-analysis-2016.pdf

Pommer M, Milliken K (2015) Pore types and pore-size distributions across thermal maturity, Eagle Ford Formation, southern Texas. AAPG Bull 99:1713–1744

Qi H, Ma J, Wong P (2002) Adsorption isotherms of fractal surfaces. Colloids Surf A Physicochem Eng Asp 206:401–407

Rigby SP (2005) Predicting surface diffusivities of molecules from equilibrium adsorption isotherms. Colloids Surf A Physicochem Eng Asp 262:139–149

Ross DJK, Bustin RM (2009) The importance of shale composition and pore structure upon gas storage potential of shale gas reservoirs. Mar Pet Geol 26:916–927

Rouquerol J, Rouquerol F, Sing KSW (1998) Absorption by powders and porous solids. Academic Press. ISBN 0080526012

Sakhaee-Pour A, Li W (2016) Fractal dimensions of shale. J Nat Gas Sci Eng 30:578–582

Schmitt M, Fernandes CP, da Cunha Neto JAB, Wolf FG, dos Santos VSS (2013) Characterization of pore systems in seal rocks using nitrogen gas adsorption combined with mercury injection capillary pressure techniques. Mar Petrol Geol 39:139–149

Shao X, Pang X, Li Q, Wang P, Chen D, Shen W, Zhao Z (2017) Pore structure and fractal characteristics of organic-rich shales: a case study of the lower Silurian Longmaxi shales in the Sichuan Basin, SW China. Mar Pet Geol 80:192–202

Sing KSW, Everett DH, Haul RAW, Moscou L, Pierotti RA, Rouquerol J, Rouquerol F, Siemieniewskat T (1985) Reporting physisorption data for gas/solid systems with special reference to the determination of surface area and porosity. Pure Appl Chem 57:603–619

Sing K (2001) The use of nitrogen adsorption for the characterisation of porous materials. Colloids Surf A 187–188:3–9

Stoeckli HF, Houriet JP (1976) The Dubinin theory of micropore filling and the adsorption of simple molecules by active carbons over a large range of temperature. Carbon 14:253–256

Stoeckli HF, Kraehenbuehl F, Ballerini L, De Bernardini S (1989) Recent developments in the Dubinin equation. Carbon 27(1):125–128

Strąpoć D, Mastalerz M, Schimmelmann A, Drobniak A, Hasenmueller NR (2010) Geochemical constraints on the origin and volume of gas in the New Albany Shale (Devonian–Mississippian), eastern Illinois Basin. AAPG Bull 94:1713–1740

Sun M, Yu B, Hu Q, Chen S, Xia W, Ye R (2016) Nanoscale pore characteristics of the Lower Cambrian Niutitang Formation Shale: a case study from Well Yuke #1 in the Southeast of Chongqing, China. Int J Coal Geol 154–155:16–29

Tian H, Pan L, Xiao X, Wilkins RWT, Meng Z, Huang B (2013) A preliminary study on the pore characterization of Lower Silurian black shales in the Chuandong Thrust Fold Belt, southwestern China using low pressure N_2 adsorption and FE-SEM methods. Mar Pet Geol 48:8–19

Thommes M, Kaneko K, Neimark AV, Oliver JP, Rodriguez-Reinoso F, Rouquerol J, Sing KSW (2015) Physisorption of gases, with special reference to the evaluation of surface area and pore size distribution (IUPAC Technical Report). In: Pure Applied Chemistry 2015. IUPAC & De Gruyter

Trunsche A (2007) Surface area and pore size determination. Mod Methods Heterogen Catal Res

Unsworth JF, Fowler CS, Jones LF (1989) Moisture in coal: 2. Maceral effects on pore structure. Fuel 68:18–26

Varma AK, Hazra B, Samad SK, Panda S, Mendhe VA (2014a) Methane sorption dynamics and hydrocarbon generation of shale samples from West Bokaro and Raniganj basins, India. J Nat Gas Sci Eng 21:1138–1147

Varma AK, Hazra B, Samad SK, Panda S, Mendhe VA, Singh S (2014b) Shale gas potential of Lower Permian shales from Raniganj and West Bokaro Basins, India. In: 66th annual meeting and symposium of the international committee for coal and organic petrology (ICCP-2014), pp 40–41

Wang FP, Reed RM (2009) Pore networks and fluid flow in gas shales. In: SPE annual technical conference and exhibition. Society of PETROLEUM Engineers, New Orleans, Louisiana, p 8. SPE 124253

Wang Y, Zhu Y, Liu S, Zhang R (2016) Pore characterization and its impact on methane adsorption capacity for organic-rich marine shales. Fuel 181:227–237

Wei M, Xiong Y, Zhang L, Li J, Peng P (2016) The effect of sample particle size on the determination of pore structure parameters in shales. Int J Coal Geol 163:177–185

Wood DA, Hazra B (2017) Characterization of organic-rich shales for petroleum exploration & exploitation: a review—part 1: bulk properties, multi-scale geometry and gas adsorption. J Earth Sci 28(5):739–757

Xia J, Song Z, Wang S, Zeng W (2017) Preliminary study of pore structure and methane sorption capacity of the Lower Cambrian shales from the north Gui-zhou Province. J Nat Gas Sci Eng 38:81–93

Xie H (1996) Fractal—an introduction to lithomechanics. Scientific Press, Beijing. 369 pp. (in Chinese)

Yang F, Ning Z, Liu H (2014) Fractal characteristics of shales from a shale gas reservoir in the Sichuan Basin, China. Fuel 115:378–384

Yang F, Ning Z, Wang Q, Liu H (2016a) Pore structure of Cambrian shales from the Sichuan Basin in China and implications to gas storage. Mar Petrol Geol 70:14–26

Yang R, He S, Yi J, Hu Q (2016b) Nano-scale pore structure and fractal dimension of organic-rich Wufeng-Longmaxi shale from Jiaoshiba area, Sichuan Basin: investigations using FE-SEM, gas adsorption and helium pycnometry. Mar Pet Geol 70:27–45

Yao Y, Liu D, Tang D, Tang S, Huang W (2008) Fractal characterization of adsorption-pores of coals from North China: an investigation on CH_4 adsorption capacity of coals. Int J Coal Geol 73:27–42

Yuan W, Pan Z, Li X, Yang Y, Zhao C, Connell LD, He J (2014) Experimental study and modelling of methane adsorption and diffusion in shale. Fuel 117:509–519

Zhang T, Ellis GE, Ruppel SC, Milliken KL, Yang R (2012) Effect of organic matter type and thermal maturity on methane adsorption in shale-gas systems. Org Geochem 47:120–131

Zhu X, Cai J, Wang X, Zhang J, Xu J (2014) Effects of organic components on the relationships between specific surface areas and organic matter in mudrocks. Int J Coal Geol 133:24–34

Chapter 8
Summary

Geochemical profiling-data of unconventional shale reservoirs is a key step in their characterization. However, it can provide misleading and ambiguous interpretation due to a lack of thorough understanding of the data and its limitations. Open-system programmed pyrolysis experiments (such as Rock-Eval) and organic petrological techniques are frequently used for geochemical screening of source rocks. The pyrolysis technique is more widely used because it is quicker, cheaper and easier to generate useful screening data with that technique. Significant insight can be gained by assessing shale reservoirs using the Rock-Eval pyrolysis technique, e.g., organic-matter richness, petroleum generation potential and thermal maturity levels.

Different kerogen types have distinct petroleum generation potential, essentially controlled by hydrogen/carbon ratio, oxygen contents, and thermal maturity levels. Rock-Eval analysis can distinguish the different kerogen types because they generate distinctive pyrogram signatures. Type I-II kerogens due to their higher hydrogen content and concomitant greater reactivity, yield more petroleum fluids under S2 curve of Rock-Eval analysis (even at lower sample weights). The S2 pyrograms for type I-II kerogens are typically associated with narrow Gaussian peak shape. In contrast, type III-IV kerogens are marked by lower petroleum fluid yields, and characteristically show right-side tailed S2 pyrograms with broader peak shapes. Close monitoring of pyrogram shapes therefore provides useful shale characterization in terms of kerogen quality.

Flame ionization detector (FID) Rock-Eval signals, although less frequently monitored in detail, can also provide useful quality information regarding organic-rich shales. As hydrogen-rich kerogens (type I and II) generate more petroleum fluids, they tend to saturate the FID signals generated, particularly as sample weights increase. FID signal saturation can lead to improper estimation of S2 peak magnitudes, hydrogen index (HI), and a general broadening of the S2 pyrogram. In turn, this results in less precise T_{max} values. On the other hand, when the S2 values are very low due to low kerogen contents or H-poor kerogen, the T_{max} values are likely to be inaccurate due to statistically insufficient FID counts. Type III-IV kerogens often present a different challenge for Rock-Eval data interpretation. While their lower S2-peak yield doesn't lead to FID saturation, the CO_2 produced from them during oxidation stage (represented by S4CO$_2$ oxidation curve) may overlap with the CO$_2$ from inorganic sources (represented by S5 curve), leading to improper TOC calculation and source rock interpretation. Keeping lower sample weights, allows complete combustion of organic-matter at lower oxidation temperatures, and thereby generates more accurate estimations of TOC. S2, S4CO$_2$, S5 peak graphics need to be critically examined and monitored to ensure meaningful Rock-Eval interpretations. Similarly, for petroleum modeling and fluid-in-place estimation, the impact of the shale matrix and inert organic-matter should be carefully monitored and corrections made, especially for type III kerogen-bearing sediments. The quantity of petroleum fluids retained in shales post generation is influenced by the mineralogy of the matrix. Clay minerals, particularly illite, tend to retain the most petroleum. Clay mineralogy and overall clay-content of organic-rich shales is important for their characterization, as some clay minerals are known to have catalytic affects on kerogen reaction kinetics in certain conditions. Moreover, as clay-rich shales tend to respond less favorably to fracture stimulation than silica-rich shales, it is possible to have shale formations with excellent petroleum generation potential but that perform poorly as reservoirs in terms of production rates achievable.

Establishing the reaction kinetics of the kerogen mixtures contained within organic-rich shales that best match the observed thermal maturity levels (e.g., measured in terms of vitrinite reflectance or other geochemical biomarkers in a suite of samples experiencing different burial histories) is essential for determining the timing and extent of petroleum generation from specific shale formation over a particular range of burial and thermal conditions. Most thermal maturity models are based on the Arrhenius equation and consider distributions of activation energies (E) at specific pre-exponential factors (A). There are benefits in calculating a cumulative time-temperature index ($\sum TTI_{ARR}$) from an Arrhenius equation integral of temperature over time involving representative single E and A values. This makes it possible to accurately model levels of thermal maturity reached by organic-rich shales linked to their burial histories over geological time scales. For quantifying the extent of conversion of a shales kerogens into petroleum a range or distribution of kerogen kinetics (i.e., several E and A combinations) is required, rather than single E and A values used for thermal maturity modelling, to take into account the various kerogen compositions present and the multiple first-order chemical reactions involved in petroleum generation. The ranges of kerogen present in organic-rich shales fall within a well-defined trend of kerogen kinetics, typically with dominant E–A pairs of values. The existence of this E–A trend makes it possible to calculate the cumulative petroleum transformation fractions applying appropriate kinetic values that are typically related to values close to that defined E–A trend. Distributions of E–A reaction kinetics also facilitate the accurate fitting of the Rock-Eval pyrolysis S2 peaks generated by shales with single or multiple kerogen types making it possible to identify the dominant reaction kinetics involved. Thermally immature shales provide more reliable estimates of their kerogen kinetics than mature samples. This is because first-order reaction kinetics associated with petroleum generation dominates the pyrolysis S2 peaks formed from thermally immature samples. In contrast, second order reactions associated with cracking of petroleum fluids already generated and non-kinetic processes (e.g. petroleum retention by micro-porosity and non-organic minerals) are also influencing the pyrolysis S2 peaks of thermally mature organic-rich shale samples.

Porous structures and pore size distributions of organic-rich shales vary significantly and, as it is within the pores that much of their petroleum resides, provide useful characterization information. Low pressure gas adsorption (LPGA) techniques are typically used to assess shale porosity. However, the interpretation of LPGA data can be ambiguous when mapping nanopores and corresponding fractal dimensions, as these are typically undercounted using nitrogen (N_2) as the adsorbate. Experiments conducted on organic-rich and organic-lean shales reveal that the porosity of organic-rich horizons is frequently underestimated when N_2 is used in LGPA techniques. However, nanopores present within the organic-matter can accessed and recorded more precisely when CO_2 is used as the adsorbate.

Open system programmed pyrolysis devices, such as Rock-Eval, require sample sizes of between 5 and 30 mg and particle crush-sizes of those samples to be approximately 212 microns, in order to generate precise and reliable results. The optimum sample size within the range mentioned will depend on the nature of the sample. Shales containing type I/II kerogens with high petroleum yielding potential, will generate large S2 peaks and are best analyzed with smaller sample sizes with the Rock-Eval equipment currently available. Lower TOC shales containing type III/IV kerogens with low H/C ratios will generate small S2 peaks and are best analyzed with larger sample sizes. On the other hand, for high-TOC type III/IV kerogen-bearing shale, sample weights should not be too high in order to avoid creating unrealistic S4CO$_2$ oxidation peaks. Notwithstanding these issues, single samples of 30 mg or less often do not provide measurements that can be considered to fully represent and characterize specific layers within a shale formation leading to uncertainty. Developing pyrolysis equipments with higher FID and infrared (IR) detector capacities would enable more precise detection of high petroleum and CO_2 yields generated during pyrolysis and enable larger sample sizes to be used in the analysis. In order to reduce uncertainties regarding how representative pyrolysis results are of formations in general, multiple samples need to be analyzed from each zone of interest. Repeating analysis on a number of sample sub-sets is also necessary to establish confidence that sample analysis is representative of the shale formation from which the samples are taken. For laterally and vertically extensive heterogeneous shale formations, multiple samples taken at different geographic locations and stratigraphic levels within the formations need to be analyzed to provide results that can be considered as representative of the formation as a whole.

The general observation that geologically younger sediments tend to display higher oxygen indices (OI) than older rocks, needs to be verified and reasoned in terms of organic-matter chemistry with changing maturity levels. While the presence of stable oxygen moieties in older mature rocks has been opined by many, a critical evaluation of the influence of such compounds and carbonate minerals on Rock-Eval S3 and S3' parameters of organic-rich shales might reveal some influences from oxygenated compounds during pyrolysis. Further studies regarding delineation of carbonate mineral species in terms of their peak-dissociation pyrolysis and oxidation temperatures; can add another dimension to the applicability of Rock-Eval for assessing unconventional shale reservoirs.

The biomarkers and its stable isotope proxies have evolved into a mature analytical method for the characterization of conventional petroleum systems. However, in gas shale systems, the cause and effect relationship between the subsurface P–T (pressure–temperature) conditions and molecular and isotopic signatures needs to be better understood. The isotopic rollovers witnessed in the shale gases from several producing basins of USA and China requires detailed investigations into the processes causing the thermal transformation of kerogen into the gas. Precise estimations of thermal maturity of kerogen in such basins are important, as they help to better constrain the petroleum generation potential of organic-rich shale formations.

LPGA studies provide key information for characterizing the porous structures in organic-rich shales. However, further refinements of the techniques and precise limitation of adsorbates used are required and much more detailed evaluations conducted on a global scale are required to confirm and generalize the currently available interpretations based on a limited set of samples, formations and geographic locations. Further, more focus should be directed towards correlating the lab-scale crushed-sample porosity observations to in situ field-scale observations, especially for CO_2 sequestration studies.

Printed in the United States
By Bookmasters